T0302135

Advanced Topics in Power Systems Analysis

Electric Power Systems Analysis is one of the most challenging courses in the Electric Power Engineering major which is taught to junior students. Its complexity arises from numerous prerequisites, a wide array of topics, and a crucial dependence on computational tools, presenting students with significant challenges.

This book serves as a continuation of our previous book, *Fundamentals of Power Systems Analysis 1: Problems and Solutions*, specifically delving into advanced topics in power systems analysis.

The structure of *Advanced Topics in Power Systems Analysis* is as follows:

Economic Load Dispatch, Symmetrical and Unsymmetrical Short Circuits, Transient Stability Analysis, Power System Linear Controls, and Key Concepts in Power System Analysis, Operation, and Control.

The structure of *Fundamentals of Power Systems Analysis 1* is as follows:

Introduction to the Power System, Transmission Line Parameters, Line Model and Performance, and Power Flow Analysis.

In brief, the advantages associated with delving into both books are as follows:

- A variety of tests to prepare for employment exams.
- Electrical engineers practicing power systems analysis can find almost everything they need.
- This book contains both difficult and easy problems and solutions.
- Readers have the capability to solve problems presented in this book solely using a calculator, without dependence on computer-based software.
- This book provides power systems concepts through studying two-choice questions.

In the end, we had a great time writing this book, and we truly hope you enjoy reading it as much as we enjoyed creating it!

Advanced Topics in Power Systems Analysis
Problems, Methods, and Solutions

Mostafa Eidiani and Kumars Rouzbehi

CRC Press
Taylor & Francis Group
Boca Raton London New York

CRC Press is an imprint of the
Taylor & Francis Group, an **informa** business

First edition published 2025
by CRC Press
2385 NW Executive Center Drive, Suite 320, Boca Raton FL 33431

and by CRC Press
4 Park Square, Milton Park, Abingdon, Oxon, OX14 4RN

CRC Press is an imprint of Taylor & Francis Group, LLC

© 2025 Mostafa Eidiani and Kumars Rouzbehi

First edition published by CRC Press 2025

ISBN: 978-1-032-82878-7 (hbk)
ISBN: 978-1-032-82866-4 (pbk)
ISBN: 978-1-003-50675-1 (ebk)
ISBN: 978-1-003-50677-5 (eBook+)

DOI: 10.1201/9781003506751

Typeset in Times
by codeMantra

Eidiani *dedicated this book to his beloveds, Elham and Hosna.*

Rouzbehi *presented this book to his late mother, who continues to be the guiding light in every page turned.*

Contents

Foreword

The outlook of power systems engineering is proof of human ingenuity, innovation, and the relentless pursuit of excellence. In the static and dynamic landscapes of electrical power systems, the pursuit of **Planning and Operation**, **Stability and Control**, and **Protection** remains an ever-evolving challenge. As the demands on electrical power systems continue to grow, so too does the need for advanced analysis techniques that can unravel complexities, anticipate issues, and offer effective solutions.

Advanced Topics in Power Systems Analysis: Problems, Methods, and Solutions is an exploration into the mentioned realms of power systems analysis and design. Through an amalgamation of theoretical insights, problems, and methodologies, this book offers a holistic view of power systems analysis. It doesn't merely scratch the surface but plunges into the heart of simple and complex scenarios, guiding readers through multifaceted challenges and providing them with the tools and methodologies to navigate them effectively. It serves as a guide, a valuable point of reference, and an essential resource for individuals aiming not only to deepen their understanding of power systems but also to attain expertise in navigating their complexities and advancing the field.

Each chapter within this book represents not just a theoretical discourse but a practical roadmap, offering insights and methodologies essential for comprehending, analyzing, and resolving problems encountered in power systems. Whether navigating the intricacies of Three-Phase Symmetrical and Unsymmetrical Short Circuits, addressing Transient Stability Analysis, or deeper understanding of Power Systems Operation and Control, this book serves as an indispensable guide.

The collaborative effort of professionals who have contributed their expertise to this volume showcases their dedication and commitment to the power systems field. Their collective insights and methods make this book a guiding light for both newcomers and experienced professionals, helping them navigate the complexities present in today's power systems. It encapsulates not just the present understanding but also paves the way for future advancements, encouraging continual learning and exploration within this dynamic field.

Thus, as you read through these chapters and learn more about advanced power systems, I hope it not only gives you useful information but also sparks your curiosity to discover more and make our electric power systems even better for the future.

Professor Hassan Modir Shanechi
Electrical and Computer Engineering Department
Illinois Institute of Technology, Chicago, USA

Preface

Electric Power Systems Analysis is one of the most challenging courses of the Electric Power Engineering major which is taught for junior students. Its complexity arises from numerous prerequisites, a wide array of topics, and a crucial dependence on computational tools, presenting students with significant challenges.

This book serves as a continuation of our previous book, "Fundamentals of Power Systems Analysis 1: Problems and Solutions", specifically delving into advanced topics in power systems analysis. Each chapter within this book acts as a stepping stone toward a deeper comprehension of the subject matter, resembling the content covered in a graduate-level course.

It is structured as follows: The first chapter "Economic Load Dispatch" gives a lesson on "power systems operation", whose prerequisite is "numerical analysis". The second and third chapters on "symmetrical and unsymmetrical short circuits", serve as an entrance door to "electrical distribution systems" and "advanced power systems protection" courses. A prerequisite for the second chapter is "Electric Circuits, I and II", and a prerequisite for the third chapter is "AC Electric Machines". The fourth chapter on "Transient Stability Analysis" introduces "Nonlinear Power System Dynamics". Prerequisites for the fourth chapter are "Engineering Mathematics" and "Differential Equations".

Power system linear controls are briefly discussed in the fifth chapter. Also, automatic voltage regulator (AVR) and automatic load frequency control (ALFC) are reviewed in this chapter. Chapter 6 takes a closer look at the subjects covered in the book. The relationship between the information in the earlier chapters and the remaining information not covered in the power systems is explained in this chapter.

This book should be considered as a supplement to the course entitled "Advanced Power Systems Analysis". Students can use this book to deepen their knowledge, improve their study speed, and review the most critical lessons of the course to prepare for the course exam.

This book is a valuable resource for both undergraduate (BEE) and graduate (MEE) students, as well as professors in the field of electrical engineering. Additionally, it serves as a convenient reference for electrical engineers working in utility companies and consulting firms, providing quick access to essential formulas and concepts.

This book contains an explanation sheet that can be used to reread and summarize needed materials to solve the related power systems problems.

In brief, the other advantages associated with delving into this book are as follows:

✓ A variety of tests to prepare for employment exams.
✓ Electrical engineers practicing power system analysis can find almost everything they need.
✓ This book contains both difficult and easy problems and solutions.

✓ Readers can solve problems presented in this book solely using a calculator, without dependence on any computer-based software.
✓ This book provides power systems concepts through studying two-choice questions.

In the end, we had a great time in the writing phase of this book, and we truly hope you enjoy reading it as much as we enjoyed creating it!

Authors

Mostafa Eidiani (StM'98, SM'16) was born in Mashhad, Iran. He earned a BS (with distinction) and an MEng. degree in electrical engineering from the Ferdowsi University of Mashhad, Iran in 1995 and 1997, respectively, as well as his PhD degree from the Science and Research Branch of Islamic Azad University, Tehran, Iran in 2004. In 2016, he was elected a senior member of IEEE, and in 2017, he was elected director of the Iranian Association of Electrical and Electronics Engineers (Khorasan Branch). He was promoted from assistant professor to associate professor in 2016. His research interests include renewable energy integration, power system control, transient and voltage stability, power system simulation, and DIgSILENT PowerFactory.

He has authored or co-authored eight technical books, nine chapter books, 40 journal papers, and 110 technical conference proceedings. In addition, he has conducted more than 30 research projects with Iranian power companies.

Dr. Eidiani is an associate editor for the *IET Journal of Engineering* (JOE) and *Journal of Electrical Engineering & Technology* (JEET), and a member of the Editorial Board for the *International Journal of Applied Power Engineering* (IJAPE) and nine other journals. He has been a board member of Khorasan Electric Generation Company. Dr. Eidiani is the editor-in-chief of the *International Journal of Energy Security and Sustainable Energy*. He is the author of "Fundamentals of Power Systems Analysis 1: Problems and Solutions" published by Taylor & Francis Group.

Kumars Rouzbehi (Senior Member, IEEE) received his PhD degree in Electric Energy Systems from the Technical University of Catalonia (UPC), Barcelona, Spain, in 2016. Prior to this, he was an academic staff at the Islamic Azad University (IAU), Iran, from 2002 to 2011. In parallel with teaching and research at the IAU, he was the CEO of Khorasan Electric and Electronics Research Company, from 2004 to 2010. From 2017 to 2018, he was an associate professor at the Loyola Andalucía University, Seville, Spain. In 2019, he joined the Department of System Engineering and Automatic Control, University of Seville, Spain. He is the patent holder for AC grid synchronization of voltage source power converters and has contributed to over 100 technical publications, including books, book chapters, journal papers, and technical conference proceedings. Professor Rouzbehi has been a TPC member of the International Conference on Electronics, Control, and Power Engineering (IEEE. ECCP); since 2014, a scientific board member of the (IEA) International Conference on Engineering and Management; and since 2015, a TPC member of COMPEL 2020.

He is an associate editor of the *IEEE Systems Journal*, *IET Generation, Transmission & Distribution*, *IET Renewable Power Generation*, *High Voltage (IET)*, and *IET Systems Integration*. He received the Second Best Paper Award 2015 from the IEEE Power Electronics Society, IEEE *Journal of Emerging and Selected Topics in Power Electronics*.

1 Economic Load Dispatch

Part One: Lesson Summary

1.1 INTRODUCTION

This chapter explains how the real power output of a controlled generating unit is selected to meet a given load and to minimize the total operating costs. This is the **economic load dispatch** or **optimal power flow problem**. Under certain conditions, minimizing the total operating cost of power plants is our main objective. This situation occurs when we have many load flow choices. The acceptable bus voltage range, which exceeds the nominal value by ±5%, makes this mode possible. Each power plant's generation cost should be calculated according to its generating power. Now, considering the importance of the problem and taking into account the transmission losses and generation limits, it is possible to solve an economic load dispatch problem.

1.2 MINIMIZATION

The simplest optimization formula:

$$\mathrm{Max}(f(x)) = \mathrm{Min}(-f(x))$$

$$f(x) = f(x_1, \cdots, x_n)$$

$$\Rightarrow \frac{\partial f}{\partial x_i} = 0, \ i = 1, 2, \ldots, n$$

Example 1.1

$$\mathrm{Min}\, f(x) = 3x_1^2 + 2x_2^2 + x_3^2 + x_2 x_1 + x_3 x_2 - 32x_1 - 16x_2 - 8x_3 + 220$$

$$
\begin{cases}
\dfrac{\partial f_1}{\partial x_1} = 6x_1 + x_2 - 32 = 0 \\[2mm]
\dfrac{\partial f_1}{\partial x_2} = 4x_2 + x_1 + x_3 - 16 = 0 \\[2mm]
\dfrac{\partial f_1}{\partial x_3} = 2x_3 + x_2 - 8 = 0
\end{cases}
$$

DOI: 10.1201/9781003506751-1

$$\begin{bmatrix} 6 & 1 & 0 \\ 1 & 4 & 1 \\ 0 & 1 & 2 \end{bmatrix} \begin{bmatrix} x_1 \\ x_2 \\ x_3 \end{bmatrix} = \begin{bmatrix} 32 \\ 16 \\ 8 \end{bmatrix} \Rightarrow \begin{bmatrix} x_1 \\ x_2 \\ x_3 \end{bmatrix}$$

$$= \begin{bmatrix} 0.175 & -0.05 & 0.025 \\ -0.05 & 0.3 & -0.15 \\ 0.025 & -0.15 & 0.575 \end{bmatrix} \begin{bmatrix} 32 \\ 16 \\ 8 \end{bmatrix} = \begin{bmatrix} 5 \\ 2 \\ 3 \end{bmatrix}$$

1.3 EQUALITY-CONSTRAINED OPTIMIZATION WITH THE LAGRANGE METHOD

$$\text{Min } f(x_1, x_2, \ldots, x_n)$$

$$\text{Subjet to: } g_j(x_1, x_2, \ldots, x_n) = 0 \tag{1.1}$$

$$j = 1, 2, \ldots, k$$

where
  $f(x)$ is Objective function
  $g(x)$ is Equality constraint

Lagrange method:

$$\text{Min } L \triangleq f + \sum_{j=1}^{k} \lambda_j g_j \tag{1.2}$$

where
  L is Lagrangian or Lagrange function
  λ is Lagrange multiplier or Lagrange coefficient

$$L = L\left(x_1, \ldots, x_n, \lambda_1, \ldots, \lambda_k\right)$$

$$\begin{cases} \dfrac{\partial L}{\partial x_i} = 0 \quad i = 1, 2, \ldots, n \\[2mm] \dfrac{\partial L}{\partial \lambda_i} = 0 \quad i = 1, 2, \ldots, k \end{cases} \Rightarrow \begin{cases} \dfrac{\partial L}{\partial x_i} = 0 = \dfrac{\partial f}{\partial x_i} + \sum_{j=1}^{k} \left(\lambda_j \dfrac{\partial g_j}{\partial x_i} \right) \quad i = 1, 2, \ldots, n \\[2mm] \dfrac{\partial L}{\partial \lambda_i} = 0 = g_i \qquad\qquad\qquad\qquad i = 1, 2, \ldots, k \end{cases} \tag{1.3}$$

1.4 INEQUALITY-CONSTRAINED OPTIMIZATION WITH THE KUHN–TUCKER METHOD

$$\text{Min } f(x_1, x_2, ..., x_n)$$

$$s.t.: \begin{cases} g_i(x_1, x_2, ..., x_n) = 0, & i = 1:k \\ \\ u_i(x_1, x_2, ..., x_n) \leq 0, & i = 1:m \end{cases} \tag{1.4}$$

where $u(x)$ is inequality constraint.

1.4.1 Kuhn–Tucker Method or Karush–Kuhn–Tucker Conditions

$$L = f + \sum_{i=1}^{k} \lambda_i g_i + \sum_{i=1}^{m} \mu_i u_i \tag{1.5}$$

$$\mu_j u_j = 0, \quad \mu_j > 0$$

$$L = L(x_1, ..., x_n, \lambda_1, ..., \lambda_k, \mu_1, ..., \mu_m)$$

$$\begin{cases} \dfrac{\partial L}{\partial x_j} = 0 = \dfrac{\partial f}{\partial x_j} + \sum_{i=1}^{k} \lambda_i \dfrac{\partial g_i}{\partial x_j} + \sum_{i=1}^{m} \mu_i \dfrac{\partial u_i}{\partial x_j} & j = 1:n \\ \\ \dfrac{\partial L}{\partial \lambda_j} = 0 = g_j & j = 1:k \\ \\ \dfrac{\partial L}{\partial \mu_j} = u_j \leq 0 & j = 1:m \end{cases} \tag{1.6}$$

1.5 ECONOMIC LOAD DISPATCH WITHOUT TRANSMISSION LOSSES AND POWER GENERATION LIMITS

The cost function of the ith power plant : $C_i(P_i) = \alpha_i + \beta_i P_i + \gamma_i P_i^2$ (1.7)

Constant parameters related to the ith power plant : $(\alpha_i, \beta_i, \gamma_i)$ (1.8)

$$\text{Total cost : } C_t = \sum_{i=1}^{ng} C_i \tag{1.9}$$

Problem description:

$$\text{Min } C_t = \sum_{i=1}^{ng} C_i(P_i)$$

$$(1.10)$$

$$s.t.: P_D = \sum_{i=1}^{ng} P_i$$

In this chapter, ng is the number of power plants participating in the economic load dispatch, and P_D is the total load demand.

Lagrange method:

$$L = C_t + \lambda \left(P_D - \sum_{i=1}^{ng} P_i \right)$$

$$(1.11)$$

$$L = L\left(P_1, P_2, \ldots, P_{ng}, \lambda\right)$$

$$\Rightarrow \begin{cases} \dfrac{\partial L}{\partial P_j} = 0 = \dfrac{\partial C_t}{\partial P_j} + \lambda(0-1) \\[4mm] \dfrac{\partial L}{\partial \lambda} = 0 = P_D - \sum_{i=1}^{ng} P_i \end{cases} \Rightarrow \dfrac{\partial C_t}{\partial P_j} = \dfrac{\partial C_j(P_j)}{\partial P_j} = \dfrac{dC_j}{dP_j} \triangleq I_{C_j} \qquad (1.12)$$

General equations:

$$\begin{cases} I_{C_j} = \lambda, \quad j = 1:ng \\[4mm] P_D = \sum_{i=1}^{ng} P_i \end{cases} \qquad I_C \text{ is incremental cost} \qquad (1.13)$$

Special case if C_i is a quadratic equation:

$$C_i(P_i) = \alpha_i + \beta_i P_i + \gamma_i P_i^2 \Rightarrow I_{Ci} = \beta_i + 2\gamma_i P_i$$

$$\Rightarrow \begin{cases} \lambda = \beta_i + 2\gamma_i P_i, \quad \Rightarrow P_i = \dfrac{\lambda - \beta_i}{2\gamma_i}, \quad i = 1:ng \\[4mm] \sum_{i=1}^{ng} P_i = P_D \end{cases} \Rightarrow \sum_{i=1}^{ng} \dfrac{\lambda - \beta_i}{2\gamma_i} = P_D \qquad (1.14)$$

$$\Rightarrow \lambda \sum_{i=1}^{ng} \frac{1}{2\gamma_i} - \sum_{i=1}^{ng} \frac{\beta_i}{2\gamma_i} = P_D \Rightarrow \lambda = \frac{P_D + \sum_{i=1}^{ng} \frac{\beta_i}{2\gamma_i}}{\sum_{i=1}^{ng} \frac{1}{2\gamma_i}}, \quad P_i = \frac{\lambda - \beta_i}{2\gamma_i} \quad (1.15)$$

Example 1.2

$$\begin{cases} C_1 = 500 + 5.5P_1 + 0.005P_1^2 \\ C_2 = 700 + 5.7P_2 + 0.007P_2^2, \ P_D = 1{,}000 \text{ MW} \\ C_3 = 900 + 5.9P_3 + 0.009P_3^2 \end{cases} \Rightarrow \begin{cases} P_1 = ? \\ P_2 = ? \\ P_3 = ? \end{cases}$$

$$\lambda = \frac{1{,}000 + \dfrac{5.5}{0.010} + \dfrac{5.7}{0.014} + \dfrac{5.9}{0.018}}{\dfrac{1}{0.010} + \dfrac{1}{0.014} + \dfrac{1}{0.018}} = 10.066 \frac{\$}{\text{MWh}}$$

$$P_1 = \frac{10.066 - 5.5}{2(0.005)} = 456.6, \ P_2 = \frac{10.066 - 5.7}{2(0.007)} = 311.86,$$

$$P_3 = \frac{10.066 - 5.9}{2(0.009)} = 231.44$$

Verification: $456.6 + 311.86 + 231.44 = 999.9 \approx 1{,}000$ MW

1.6 ECONOMIC LOAD DISPATCH WITHOUT TRANSMISSION LOSSES AND WITH POWER GENERATION LIMITS

Problem:

$$\begin{cases} \text{Min } C_t = \sum_{i=1}^{ng} C_i(P_i) \\ \\ s.t : \begin{cases} \sum_{i=1}^{ng} P_i = P_D \\ \\ P_i^{min} \leq P_i \leq P_i^{max} \end{cases} \end{cases}$$

Kuhn–Tucker Method:

$$\begin{cases} \sum_{i=1}^{ng} P_i = P_D \\[2mm] \lambda = I_{C_i} \rightarrow P_i^{\min} \leq P_i \leq P_i^{\max} \\[2mm] \lambda > I_{C_i} \rightarrow P_i = P_i^{\max} \\[2mm] \lambda < I_{C_i} \rightarrow P_i = P_i^{\min} \end{cases}$$

Solving issues:

1. Finding a solution without a generation limit.
2. Verify whether or not they are out of range.
3. Out-of-range units impact power consumption because they are not part of economic load dispatch. Then economic load dispatch is done for the others and the Kuhn–Tucker condition is checked.

1.6.1 LAMBDA ITERATION TECHNIQUE

Initial guess: $\lambda^{(0)}$

$$\lambda^{(0)} \rightarrow P_i = \frac{\lambda^{(0)} - \beta_i}{2\gamma_i} \Rightarrow \begin{cases} a) \ \sum P_i < P_D \Rightarrow \lambda^{(1)} = \lambda^{(0)} + \Delta\lambda \\[2mm] b) \ \sum P_i > P_D \Rightarrow \lambda^{(1)} = \lambda^{(0)} - \Delta\lambda \end{cases}$$

Should the aforementioned process bounce between stages (a) and (b), $\Delta\lambda$s value will be halved. The aforementioned procedure keeps running until the $\Delta\lambda$ becomes negligible ($<\varepsilon$).

1.7 ECONOMIC LOAD DISPATCH WITH TRANSMISSION LOSSES AND WITH POWER GENERATION LIMITS

Loss function:

$$P_L = P_L\left(P_1, P_2, \ldots, P_{ng}\right)$$

Kron's loss formula:

$$P_L = \sum_{i=1}^{ng}\sum_{j=1}^{ng} P_i B_{ij} P_j + \sum_{i=1}^{ng} B_{0i} P_i + B_{00}$$

B_{ij} are Kron's coefficients or B coefficients.

Problem:

$$
\left\{
\begin{array}{l}
\text{Min } C_t = \displaystyle\sum_{i=1}^{ng} C_i(P_i) \\[4mm]
s.t : \left\{
\begin{array}{l}
\displaystyle\sum_{i=1}^{ng} P_i = P_D + P_L \\[4mm]
P_i^{\min} \le P_i \le P_i^{\max}
\end{array}
\right.
\end{array}
\right.
$$

Kuhn–Tucker Method:

$$
L = L\left(P_1,\ldots,P_{ng},\lambda,\mu_1^{\min},\ldots,\mu_{ng}^{\min},\mu_1^{\max},\ldots,\mu_{ng}^{\max}\right)
$$

$$
\text{Min } L = C_t + \lambda\left(P_D + P_L - \sum_{i=1}^{ng} P_i\right) + \sum_{i=1}^{ng}\mu_i^{\max}\left(P_i - P_i^{\max}\right) + \sum_{i=1}^{ng}\mu_i^{\min}\left(P_i^{\min} - P_i\right)
$$

Without requiring inequality, the problem is resolved. Finally, the inequality criteria are put into practice.

$$
L = C_t + \lambda\left(P_D + P_L - \sum_{j=1}^{ng} P_j\right); \quad L = L\left(P_1,\ldots,P_{ng},\lambda\right)
$$

$$
\left\{
\begin{array}{l}
\dfrac{\partial L}{\partial P_i} = \dfrac{\partial C_t}{\partial P_i} + \lambda\left(0 + \dfrac{\partial P_L}{\partial P_i} - 1\right) = 0 \\[5mm]
\dfrac{\partial L}{\partial \lambda} = 0 = P_D + P_L - \displaystyle\sum_{i=1}^{ng} P_i = 0
\end{array}
\right.
\Rightarrow \dfrac{\partial C_t}{\partial P_i} = \dfrac{\partial C_i}{\partial P_i} = \dfrac{dC_i}{dP_i} = I_{C_i} = \beta_i + 2\gamma_i P_i
$$

Incremental Transmission Loss:

$$
ITL_i = \frac{\partial P_L}{\partial P_i}
$$

$$
\text{if}: P_L = \sum_{i=1}^{ng}\sum_{j=1}^{ng} P_i B_{ij} P_j + \sum_{i=1}^{ng} B_{0i} P_i + B_{00} \Rightarrow \frac{\partial P_L}{\partial P_k} = 2\sum_{j=1}^{ng} B_{kj} P_j + B_{0k}
$$

Economic load dispatch conditions with transmission losses and without power generation limits:

$$\begin{cases} \lambda = \dfrac{I_{C_i}}{1 - ITL_i} = L_i I_{C_i} \\ \\ P_D + P_L = \displaystyle\sum_{i=1}^{ng} P_i \end{cases} \quad , \textbf{Penalty Factor}: L_i \triangleq \dfrac{1}{1 - ITL_i}$$

The production cost should be higher for a unit that makes more losses.

1.8 VECTOR/MATRIX RELATION TO SOLVING ECONOMIC LOAD DISPATCH PROBLEM

$$\begin{cases} \lambda = L_i I_{C_i} \\ \\ \displaystyle\sum_{i=1}^{ng} P_i = P_D + P_L \end{cases}$$

$$\lambda(1 - ITL_i) = \beta_i + 2\gamma_i P_i$$

$$\Rightarrow \lambda\left(1 - 2\sum_{j=1}^{ng} B_{ij} P_j - B_{0i}\right) = \beta_i + 2\gamma_i P_i$$

$$\lambda(1 - B_{0i}) - \beta_i = 2\gamma_i P_i + 2\lambda\sum_{j=1}^{ng} B_{ij} P_j = 2\left(\gamma_i P_i + \lambda B_{ii} P_i + \lambda\sum_{j=1\neq1}^{ng} B_{ij} P_j\right)$$

$$\begin{bmatrix} \dfrac{\gamma_1}{\lambda} + B_{11} & B_{12} & B_{13} & \cdots & B_{1ng} \\ B_{21} & \dfrac{\gamma_2}{\lambda} + B_{22} & \cdot & \cdots & B_{2ng} \\ \cdot & \cdot & \cdot & \cdot & \cdot \\ \cdot & \cdot & \cdot & \cdot & \cdot \\ B_{ng1} & B_{ng2} & \cdot & \cdots & \dfrac{\gamma_{ng}}{\lambda} + B_{ng.ng} \end{bmatrix} \begin{bmatrix} P_1 \\ P_2 \\ \cdot \\ \cdot \\ \cdot \\ P_{ng} \end{bmatrix} = \dfrac{1}{2} \begin{bmatrix} 1 - B_{01} - \dfrac{\beta_1}{\lambda} \\ 1 - B_{02} - \dfrac{\beta_2}{\lambda} \\ \cdot \\ \cdot \\ 1 - B_{0ng} - \dfrac{\beta_{ng}}{\lambda} \end{bmatrix}$$

1.9 DETERMINING COST FUNCTION PARAMETERS FOR A POWER PLANT (CURVE FITTING)

n points are known:

$$(C_1, P_1), \cdots, (C_n, P_n)$$

Total error:

$$E_r = \sum_{i=1}^{n} (C_i - C_i(P_i))^2$$

Objective function:

$$\text{Min } E_r(\alpha, \beta, \gamma)$$

$$\Rightarrow E_r = \sum_{i=1}^{n} \left(C_i - \alpha - \beta P_i - \gamma P_i^2\right)^2$$

If we define simplicity in the display:

$$\sum_{i=1}^{n} () \triangleq \sum \quad \Rightarrow$$

$$\frac{\partial E_r}{\partial \alpha} = 0 = -2\sum_{i=1}^{n}\left(C_i - \alpha - \beta P_i - \gamma P_i^2\right) \Rightarrow \sum C_i - n\alpha - \beta \sum P_i - \gamma \sum P_i^2 = 0$$

$$\frac{\partial E_r}{\partial \beta} = 0 = \sum_{i=1}^{n}(-2P_i)\left(C_i - \alpha - \beta P_i - \gamma P_i^2\right) \Rightarrow$$

$$\Rightarrow \sum C_i P_i - \alpha \sum P_i - \beta \sum P_i^2 - \gamma \sum P_i^3 = 0$$

$$\frac{\partial E_r}{\partial \gamma} = 0 = \sum_{i=1}^{n} -2P_i^2\left(C_i - \alpha - \beta P_i - \gamma P_i^2\right) \Rightarrow$$

$$\Rightarrow \sum C_i P_i^2 - \alpha \sum P_i^2 - \beta \sum P_i^3 - \gamma \sum P_i^4 = 0$$

$$\Rightarrow \begin{bmatrix} \alpha \\ \beta \\ \gamma \end{bmatrix} = \begin{bmatrix} n & \sum P_i & \sum P_i^2 \\ \sum P_i & \sum P_i^2 & \sum P_i^3 \\ \sum P_i^2 & \sum P_i^3 & \sum P_i^4 \end{bmatrix}^{-1} \begin{bmatrix} \sum C_i \\ \sum C_i P_i \\ \sum C_i P_i^2 \end{bmatrix}$$

Part Two: Answer-Question

1.10   TWO-CHOICE QUESTIONS (YES/NO)

1. "Optimal load flow" is the same as "economic load dispatch".
2. "Economic load dispatch" means determining the minimum generation cost.
3. "Economic load dispatch" can be solved with equal constraints by the Lagrange method.
4. In maximizing social welfare, the producer benefit is also being considered.
5. The Lagrange method is simpler than an iterative method.
6. In the iterative method, the initial guess is vital.
7. If the sum of generation minimums is less than the total load, "economic load dispatch" has the answer.
8. In the iterative lambda method, if the sum of generation is greater than the load, lambda should be increased.
9. Does the nuclear power plant participate in economic load dispatch?
10. The penalty factor is higher for power plants with a higher generation cost.
11. In "economic load dispatch", the losses function is a function of generating power.
12. The power plant that is closer to the load produces fewer losses than the one that is further away.
13. If the cost function is a polynomial function of degree 3, Lagrange's method can be used.
14. The cost function is usually a polynomial function of degree 2.
15. Environment and pollution affect optimal load flow.
16. Power plants always have equal incremental costs in economic load dispatch.
17. In economic load dispatch, for the unit in $P_{\min}$, lambda is less than I_c.
18. In economic load dispatch, for the unit in $P_{\max}$, lambda is less than I_c.
19. The cost function's α coefficient plays a significant role in "economic load dispatch".
20. Can the "war zone" effect be applied in economic load dispatch calculations?
21. Can the "hydropower plants" effect be applied in economic load dispatch calculations?
22. The total cost rises by lambda if the total load increases by one unit.
23. If all power plants produce at the same rate, production costs will undoubtedly rise.

1.11   ANSWERS TO TWO-CHOICE QUESTIONS

1,7,8,9,10,16,18,19	No
The rest	Yes

1.12   DESCRIPTIVE QUESTIONS OF ECONOMIC LOAD DISPATCH

1.1   Find the optimal load dispatch when the total load is 400 MW.
Difficulty level o Easy •Normal o Hard

$$P_{G1} = -100 + 50(I_{C_1}) - 2(I_{C_1})^2$$

$$P_{G2} = -150 + 60(I_{C_2}) - 2.5(I_{C_2})^2$$

$$P_{G3} = -80 + 40(I_{C_3}) - 1.8(I_{C_3})^2$$

1.2   Repeat question (1.1) with $P_L = 0.004P_1$.
Difficulty level o Easy •Normal o Hard

1.3   Find the optimal load dispatch when the total load is 360 MW.
Difficulty level o Easy •Normal o Hard

$$I_{C_1} = 9.3 + 0.041P_{G_1} + 0.00012P_{G_1}^2$$

$$I_{C_2} = 7.6 + 0.031P_{G_2} + 0.00008P_{G_2}^2$$

1.4   For (n) power plants, if ($I_{C_i} = \beta_i + 2\gamma_i P_{G_i}$, $i = 1:n$), show that in optimal load dispatch we have:
Difficulty level • Easy o Normal o Hard

$$\lambda = \frac{P_D + \sum_{i=1}^{n} \dfrac{\beta_i}{2\gamma_i}}{\sum_{i=1}^{n} \dfrac{1}{2\gamma_i}}$$

1.5   In optimal load dispatch, if the total load of the system slightly changes (ΔP_D), the total cost also changes slightly and is equal to ($\Delta C = \lambda \, \Delta P_D$).
Difficulty level • Easy o Normal o Hard

1.6   Find the optimal load dispatch when the total load is 600 MW.
Difficulty level o Easy •Normal o Hard

$$I_{C_1} = 0.012P_{G_1} + 8 \qquad 100 \le P_{G_1} \le 650 \text{ MW}$$
$$I_{C_2} = 0.018P_{G_2} + 7' \qquad 50 \le P_{G_2} \le 500 \text{ MW}$$

1.7   In question (1.6), if ($P_D = 601$ MW), find the cost.
Difficulty level o Easy •Normal o Hard

1.8 Find the optimal load dispatch when the total load is 600 MW.
Difficulty level • Easy o Normal o Hard

$$I_{C_1} = 0.019 P_{G_1} + 9 \qquad 100 \le P_{G_1} \le 250 \text{ MW}$$
$$I_{C_2} = 0.011 P_{G_2} + 6' \qquad 50 \le P_{G_2} \le 300 \text{ MW}$$

1.9 Find the optimal load dispatch.
Difficulty level o Easy o Normal • Hard

$$C_1 = 16,000 + 1,600 P_{G_1} + 4.8 P_{G_1}^2 \qquad 10 \le P_{G_i} \le 100 \text{ MW}$$

$$C_2 = 2,400 + 1,200 P_{G_2} + 8 P_{G_2}^2 \qquad 50 \le P_D \le 200 \text{ MW}$$

1.10 Find the optimal load dispatch.
Difficulty level o Easy •Normal o Hard

$$C_1 = \frac{1}{200} P_{G_1}{}^2 + 8.5 P_{G_1} + 200 \qquad \frac{\$}{hr}$$

$$C_2 = \frac{3}{400} P_{G_2}{}^2 + 9.5 P_{G_2} + 400 \qquad \frac{\$}{hr}$$

$$\lambda = 13.5, \qquad P_L = 0.0004 P_{G_1}^2 \quad \text{MW}$$

1.11 Find the optimal load dispatch when the total load is 975 MW.
Difficulty level o Easy o Normal • Hard

$$\begin{cases} 200 \le P_1 \le 450 \\ 150 \le P_2 \le 350 \\ 190 \le P_3 \le 300 \end{cases} \begin{cases} C_1 = 500 + 5.3 P_1 + 0.004 P_1^2 \\ C_2 = 400 + 5.5 P_2 + 0.006 P_2^2 \\ C_3 = 200 + 5.8 P_3 + 0.009 P_3^2 \end{cases}$$

1.13 DESCRIPTIVE ANSWERS TO ECONOMIC LOAD DISPATCH

1.1 Under optimal dispatch conditions, we have:

$$I_{C1} = I_{C2} = I_{C3} \triangleq I_C \text{ and } P_{G_1} + P_{G_2} + P_{G_3} = P_D$$

Then:

$$\Rightarrow -100 + 50 I_C - 2 I_C^2 - 150 + 60 I_C - 2.5 I_C^2 - 80 + 40 I_C - 1.8 I_C^2 = 400$$

$$\Rightarrow -330 + 150 I_C - 6.3 I_C^2 = 400 \Rightarrow 6.3 I_C^2 - 150 I_C + 730 = 0 \Rightarrow \begin{cases} I_C = 16.99 \\ I_C = 6.82 \end{cases}$$

$$\text{If}: I_C = 16.99 \Rightarrow \begin{cases} P_{G_1} = 172.1798 \\ P_{G_2} = 147.74 \\ P_{G_3} = 80.0118 \end{cases} , \text{If}: I_C = 6.82 \Rightarrow \begin{cases} P_{G_1} = 147.98 \\ P_{G_2} = 142.92 \\ P_{G_3} = 109.08 \end{cases}$$

1.2 $\begin{cases} \lambda = L_i I_{C_i} \\ P_1 + P_2 + P_3 = P_D + P_L \end{cases}$

$$L_1 = \frac{1}{1 - ITL_1} = \frac{1}{1 - 0.004} = \frac{1}{0.996}, \; L_2 = L_3 = 1$$

$$\Rightarrow \begin{cases} \lambda = \dfrac{I_{C_1}}{0.996} = I_{C_2} = I_{C_3} \\ P_1 + P_2 + P_3 = 400 + 0.004 P_1 \end{cases} \Rightarrow \begin{cases} I_{C_1} = 0.996\lambda, \; I_{C_2} = I_{C_3} = \lambda \\ 0.996 P_1 + P_2 + P_3 = 400 \Rightarrow \end{cases}$$

$$\begin{cases} \Rightarrow 0.996\left(-100 + 50(0.996\lambda) - 2(0.996\lambda)^2\right) + \\ (-150 + 60\lambda - 2.5\lambda^2) + (-80 + 40\lambda - 1.8\lambda^2) = 400 \end{cases}$$

$$\Rightarrow -729.6 + 149.6\lambda - 6.28\lambda^2 = 0$$

$$\Rightarrow \lambda = \begin{cases} 6.84 \Rightarrow P_1 = 147.8, \; P_2 = 143.4, \; P_3 = 109.4 \quad \text{MW} \\ 16.98 : \text{unacceptable?} \end{cases}$$

1.3 Under optimal dispatch conditions, we have:

$$I_{C1} = I_{C2} \triangleq I_C \text{ and } P_{G_1} + P_{G_2} = P_D$$

Then:

$$\begin{cases} 9.3 + 0.041 P_{G_1} + 0.00012 P_{G_1}^2 = 7.6 + 0.031 P_{G_2} + 0.00008 P_{G_2}^2 \\ P_{G_1} + P_{G_2} = 360 \Rightarrow P_{G_1} = 360 - P_{G_2} \end{cases}$$

$$P_{G_2} = \begin{cases} 213.62 \Rightarrow \\ 3,746.38 > P_D : (\text{unacceptable}) \end{cases} \Rightarrow P_{G_1} = 360 - 213.62 = 146.38$$

1.4   Under optimal dispatch conditions, we have:

$$\lambda = I_{C_i} = \beta_i + 2\gamma_i P_{G_i} \Rightarrow \frac{\lambda}{2\gamma_i} = \frac{\beta_i}{2\gamma_i} + P_{G_i} \Rightarrow \sum_{i=1}^{n}\frac{\lambda}{2\gamma_i} = \sum_{i=1}^{n}\frac{\beta_i}{2\gamma_i} + \sum_{i=1}^{n} P_{G_i}$$

$$\Rightarrow \lambda \sum_{i=1}^{n}\frac{1}{2\gamma_i} = \sum_{i=1}^{n}\frac{\beta_i}{2\gamma_i} + P_D \Rightarrow \lambda = \frac{\displaystyle\sum_{i=1}^{n}\frac{\beta_i}{2\gamma_i} + P_D}{\displaystyle\sum_{i=1}^{n}\frac{1}{2\gamma_i}}$$

1.5   Under optimal dispatch conditions, we have:

$$\lambda = I_{C_i} = \frac{dC_i}{dP_{G_i}} \Rightarrow \lambda\, dP_{G_i} = dC_i \Rightarrow \sum_{i=1}^{n}\lambda\, dP_{G_i} = \sum_{i=1}^{n} dC_i$$

$$\Rightarrow \lambda \sum_{i=1}^{n} dP_{G_i} = \sum_{i=1}^{n} dC_i \xrightarrow{\Delta P_{gi}\approx 0} \lambda \sum_{i=1}^{n}\Delta P_{G_i} = \sum_{i=1}^{n}\Delta C_i \xrightarrow{P_L=0} \lambda\,\Delta P_D = \Delta C$$

1.6   Under optimal dispatch conditions, we have:

$$I_{C_i} = \lambda \Rightarrow \begin{cases} 0.012P_{G_1} + 8 = 0.018P_{G_2} + 7 \\[2mm] P_{G_1} + P_{G_2} = 600 \end{cases}$$

$$\Rightarrow \begin{cases} 1.8P_{G_2} - 1.2P_{G_1} = 100 \\[2mm] P_{G_2} + P_{G_1} = 600 \end{cases} \Rightarrow \begin{cases} P_{G_1} = 326.667 \ \text{MW} \\[2mm] P_{G_2} = 273.333 \ \text{MW} \end{cases}$$

$$\Rightarrow \lambda = (0.012)(326.667) + 8 = 11.92$$

Or from question 1.4 we have:

$$\lambda = \frac{\dfrac{8}{0.012} + \dfrac{7}{0.018} + 600}{\dfrac{1}{0.012} + \dfrac{1}{0.018}} = 11.92 \Rightarrow \begin{cases} P_{G_1} = \dfrac{11.92-8}{0.012} = 326.667 \ \text{MW} \\[4mm] P_{G_2} = \dfrac{11.92-7}{0.018} = 273.333 \ \text{MW} \end{cases}$$

1.7

First solution: From answer 1.6, we have:

$$\lambda = \frac{\dfrac{8}{0.012} + \dfrac{7}{0.018} + 601}{\dfrac{1}{0.012} + \dfrac{1}{0.018}} = 11.927 \Rightarrow \begin{cases} P_{G_1} = \dfrac{11.927-8}{0.012} = 327.250 \ \text{MW} \\[2mm] P_{G_2} = \dfrac{11.927-7}{0.018} = 273.722 \ \text{MW} \end{cases}$$

$$C_i = \int I_{C_i} \, dP_{G_i} \rightarrow \begin{cases} C_1 = 0.006 P_{G_1}^2 + 8 P_{G_1} + \alpha_1 \\[2mm] C_2 = 0.009 P_{G_2}^2 + 7 P_{G_2} + \alpha_2 \end{cases} \Rightarrow \Delta C = C_{\Pi} - C_{\mathrm{I}} \Rightarrow$$

$$\Delta C = C_1(327.250) + C_2(273.722) - C_1(326.667) - C_2(273.333) \approx 11.92$$

Second solution: From answer 1.5, we have:

$$\Delta C = \lambda \, \Delta P_D = (11.92)(1) = 11.92$$

1.8 Under optimal dispatch conditions, we have:

$$\sum P_{i\min} \le P_D \le \sum P_{i\max} \Rightarrow (100+50) \le 600 \le (250+300)$$

$\Rightarrow 150 \le 600 \le 550 \Rightarrow$ Therefore, this is an impossible case of optimal dispatch.

1.9 $\lambda = I_{C_1} = I_{C_2} \Rightarrow 1{,}600 + 9.6 P_{G_1} = 1{,}200 + 16 P_{G_2} \Rightarrow 9.6 P_{G_1} - 16 P_{G_2} = -400$

and : $P_{G_1} + P_{G_2} = P_D \Rightarrow$

$$\begin{cases} 9.6 P_{G_1} - 16 P_{G_2} = -400 \\ P_{G_1} + P_{G_2} = P_D \end{cases} \Rightarrow \begin{cases} (1) : 9.6 P_{G_1} - 16 P_{G_2} = -400 \\ (2) : 16 P_{G_1} + 16 P_{G_2} = 16 P_D \end{cases}$$

$$\overset{(1)+(2)}{\Rightarrow} 25.6 P_{G_1} = 16 P_D - 400 \Rightarrow P_{G_1} = \frac{16 P_D - 400}{25.6}, P_{G_2} = \frac{9.6 P_D + 400}{25.6}$$

$$P_D = 50 \Rightarrow \begin{cases} P_{G_1} = 15.625 \\ P_{G_2} = 34.375 \end{cases}, \quad \text{Check} : 10 \ \text{MW} < P_{G_i} < 100 \ \text{MW}$$

$$P_D = 200 \Rightarrow \begin{cases} P_{G_1} = 109.375 \\ P_{G_2} = 90.625 \end{cases}, \quad \text{Check} : P_{G_1} > 100 \ \text{MW}$$

$$\Rightarrow P_{G_1} = P_{G_1 Max} = 100 \Rightarrow P_{G_1} = \frac{16P_D - 400}{25.6} \Rightarrow 100 = \frac{16P_D - 400}{25.6} \Rightarrow$$

$$\Rightarrow P_D = 185 \Rightarrow P_{G_1} + P_{G_2} = 185 \Rightarrow P_{G_2} = 85 \text{ MW}$$

Finally:

$$\Rightarrow \begin{cases} 50 \le P_D \le 185 \Rightarrow P_{G_1} = \dfrac{16P_D - 400}{25.6}, P_{G_2} = \dfrac{9.6P_D + 400}{25.6} \\ 185 \le P_D \le 200 \Rightarrow P_{G_1} = 100,\ P_{G_2} = P_D - P_{G_1} \end{cases}$$

1.10 $\lambda = \dfrac{I_{C_i}}{1 - ITL_i} = 13.5$

$$I_{C_1} = \frac{dC_1}{dP_{G_1}} = \frac{1}{100} P_{G_1} + 8.5$$

$$I_{C_2} = \frac{dC_2}{dP_{G_2}} = \frac{3}{200} P_{G_2} + 9.5$$

$$ITL_1 = \frac{\partial P_L}{\partial P_{G_1}} = 0.0008 P_{G_1}, \ \ ITL_2 = 0$$

$$\begin{cases} 13.5 = \dfrac{0.01 P_{G_1} + 8.5}{1 - 0.0008 P_{G_1}} \\ 13.5 = \dfrac{3}{200} P_{G_2} + 9.5 \end{cases} \Rightarrow \begin{cases} P_{G_1} = 240.38 \\ P_{G_2} = 266.67 \end{cases} \Rightarrow \begin{cases} P_L = 23.113 \text{ MW} \\ P_D = 483.94 \text{ MW} \end{cases}$$

1.11   Finding a solution without limiting production.
      Step 1:

$$\begin{cases} \lambda = I_{C_1} = I_{C_2} = I_{C_3} \\ P_1 + P_2 + P_3 = 975 \end{cases} \Rightarrow \lambda = 5.3 + 0.008 P_1 = 5.5 + 0.012 P_2 = 5.8 + 0.018 P_3$$

$$\begin{cases} P_1 = \dfrac{\lambda - 5.3}{0.008} \\ P_2 = \dfrac{\lambda - 5.5}{0.012} \\ P_3 = \dfrac{\lambda - 5.8}{0.018} \end{cases} \Rightarrow P_1 + P_2 + P_3 = 975 \text{ MW}$$

$$\Rightarrow \lambda = \frac{975 + \dfrac{5.3}{0.008} + \dfrac{5.5}{0.012} + \dfrac{5.8}{0.018}}{\dfrac{1}{0.008} + \dfrac{1}{0.012} + \dfrac{1}{0.018}} = 9.163$$

$$\begin{cases} P_1 = 482.89 > 450 \Rightarrow P_1 = 450 \\ P_2 = 305.26 \\ P_3 = 186.84 < 190 \Rightarrow P_3 = 190 \end{cases}$$

$$\Rightarrow P_D^{\text{New}} = 975 - 450 - 190 = 335 \Rightarrow P_2 = 335 \text{ MW}$$

$$\begin{cases} P_1 = 450 : \text{Max} \Rightarrow I_{C_1} = 8.9 \\ P_2 = 335 \\ P_3 = 190 : \text{Min} \Rightarrow I_{C_3} = 9.22 \end{cases} \Rightarrow \lambda = I_{C_2} = 5.5 + (0.012)(335) = 9.52$$

Kuhn–Tucker's condition:

$$1.\ \lambda > I_{C_1},\ 2.\ \lambda \nless I_{C_3}$$

Step 2:

The third unit has not reached the limit because the second condition is false.

$$P_1 = 450 \text{ MW}$$

$$P_D^{\text{New}} = 975 - 450 = 525 \text{ MW}$$

$$\begin{cases} P_2 + P_3 = 525 \\ \lambda = I_{C_2} = I_{C_3} \end{cases} \Rightarrow \lambda = \frac{525 + \dfrac{5.5}{0.012} + \dfrac{5.8}{0.018}}{\dfrac{1}{0.012} + \dfrac{1}{0.018}} = 9.4$$

$$\Rightarrow \begin{cases} P_2 = 325 : \lambda = 9.4 \\ P_3 = 200 : \lambda = 9.4 \\ P_1 = 450 : \text{Max} : I_{C_1} = 8.9 \end{cases} \Rightarrow \lambda > I_{C_1} \Rightarrow 9.4 > 8.9$$

Since Kuhn–Tucker's condition holds true, a solution has been found.

2 Three-Phase Symmetrical Short Circuit

Part One: Lesson Summary

2.1 INTRODUCTION

Short circuits (sorted by frequency of occurrence) happen as the following: single line-to-ground, line-to-line, double line-to-ground, and balanced three-phase faults. The path of the fault current either has zero impedance, which is named "bolted short circuit" or "nonzero impedance". Other types of faults are one-conductor-open and two-conductors-open, which can happen when conductors break or when one or two phases of a circuit breaker inadvertently open.

This chapter presents the calculations of a three-phase symmetrical short-circuit (fault) current. Even though it has been demonstrated that calculations for short circuits can be done using the regular load flow. However, a faster approach to doing short-circuit calculations is using Thévenin's method. The direct method for calculating the impedance matrix is described in the following.

In this chapter, these presumptions are used in all of the calculations: Due to the low impedance of the short circuit, the short-circuit current is much higher than the load current. As a result, loads are usually omitted when performing short-circuit calculations. The resistance of the network elements is also ignored, so the pre-fault voltage of all buses becomes $1\angle 0$.

2.2 DIRECT SOLUTION METHOD (POWER FLOW)

An example is provided to review this section. See Figure 2.1.
Short-circuit impedance in bus 3: $Z_f = j0.25$

Example 2.1

See Figure 2.2. Determine the bus voltage, short-circuit current, line, and generator current during the fault.

First, we convert the Delta impedance to Star impedance (Δ-Y transformation). See Figure 2.2. KCL at point (star center) is as follows.

$$\text{KCL}: \frac{\hat{V}_S}{j0.57} + \frac{\hat{V}_S - 1\angle 0}{j0.46} + \frac{\hat{V}_S - 1\angle 0}{j0.46} = 0$$

DOI: 10.1201/9781003506751-2

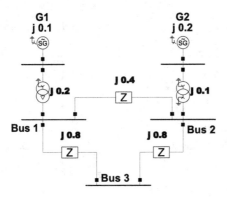

FIGURE 2.1   Impedance diagram of a simple power system.

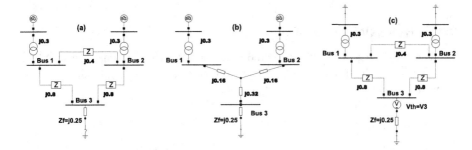

FIGURE 2.2   (a) Impedance network for fault in bus 3, (b) Equivalent network for direct solution, (c) Equivalent Thévenin network.

$$\hat{V}_S = \cfrac{\cfrac{1}{j0.46} + \cfrac{1}{j0.46}}{\cfrac{1}{j0.57} + \cfrac{1}{j0.46} + \cfrac{1}{j0.46}} = 0.7125 \Rightarrow \begin{cases} \hat{I}_{g1}^f = \dfrac{1 - \hat{V}_S}{j0.46} = -j0.625 \\[3mm] \hat{I}_{g2}^f = \dfrac{1 - \hat{V}_S}{j0.46} = -j0.625 \\[3mm] \hat{I}_3^f = \hat{I}_f = \dfrac{\hat{V}_S}{j0.57} = -j1.25 \end{cases}$$

$$\Rightarrow \begin{cases} \hat{V}_1^f = -j0.3\hat{I}_{g1}^f + 1\angle 0 = 0.8125\angle 0 \\[3mm] \hat{V}_2^f = -j0.3\hat{I}_{g2}^f + 1\angle 0 = 0.8125\angle 0 \Rightarrow \hat{I}_{12}^f = \dfrac{\hat{V}_1^f - \hat{V}_2^f}{j0.4} = 0 \\[3mm] \hat{V}_3^f = j0.25\hat{I}_f = 0.3125\angle 0 \end{cases}$$

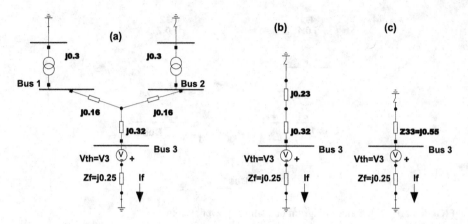

FIGURE 2.3   The reduction steps of Thévenin's equivalent circuit.

2.3   THÉVENIN METHOD

The voltage sources in Thévenin's method become zero. The fault impedance and the voltage of the faulty bus prior to the fault are in series. Figure 2.3 shows the reduction steps of Thévenin's equivalent circuit. We have:

$$
\begin{cases} \hat{V}^f = \hat{V}^0 + \hat{V}^{th} \\ \hat{I}^f = \hat{I}^0 + \hat{I}^{th} \end{cases} \quad \text{and} \quad \begin{cases} \hat{V}^0 = 1\angle 0 \\ \qquad\qquad \text{(old or pre-fault system)} \\ \hat{I}^0 = 0 \qquad\qquad \text{(0, o or O)} \end{cases} \tag{2.1}
$$

$\hat{V}^f, \hat{I}^f$ represent the faulted system's voltage and current. $\hat{V}^{th}, \hat{I}^{th}$ represent the Thévenin system's voltage and current.

KVL and from equation 2.1 we have:

$$
\hat{I}_3^f = \hat{I}_3^{th} = \frac{1}{j0.25 + j0.55} = -j1.25
$$

Current divider:

$$
\hat{I}_{g1}^{th} = \frac{j0.46}{j0.46 + j0.46}(-j1.25) = -j0.625 = \hat{I}_{g2}^{th}
$$

$$
\hat{I}_3^0 = 0 \Rightarrow \hat{I}_3^f = \hat{I}_3^{th} + \hat{I}_3^0 = -j1.25
$$
$$
\hat{I}_{g1}^0 = 0 \Rightarrow \hat{I}_{g1}^f = \hat{I}_{g1}^{th} + \hat{I}_{g1}^0 = -j0.625
$$
(it is the same as the previous method)

2.4   SHORT-CIRCUIT CAPACITY (SCC)

$$\text{SCC} = \left|\text{Pre-fault bus voltage}\right| \times \left|\text{Short-circuit current}\right| \qquad (2.2)$$

$$\text{In perunit}: \text{SCC}_k = 1 \times \left|\hat{I}_f\right| = \frac{1}{\left|Z_f + Z_{kk}\right|} \qquad (2.3)$$

For Example 2.1:

$$\text{SCC}_3 = \left|\frac{1}{j0.25 + j0.55}\right| = 1.25$$

2.5   MATRIX RELATIONS OF SHORT-CIRCUIT CALCULATIONS

Pre-fault bus voltages, known parameters, and usually $1\angle0$.

K is a short-circuited bus number, and we define: $\hat{I}_k^{(f)} \triangleq \hat{I}_f$

$$V_{\text{bus}}^{(o)} = \begin{bmatrix} \hat{V}_1^{(o)} & \cdots & \hat{V}_k^{(o)} & \cdots & \hat{V}_n^{(o)} \end{bmatrix}^T \qquad (2.4)$$

Post-fault bus voltages:

$$V_{\text{bus}}^{(f)} = \begin{bmatrix} \hat{V}_1^{(f)} & \cdots & \hat{V}_k^{(f)} & \cdots & \hat{V}_n^{(f)} \end{bmatrix}^T \qquad (2.5)$$

We have (the Thévenin method):

$$V_{\text{bus}}^{(f)} = V_{\text{bus}}^{(o)} + V_{\text{bus}}^{(th)} \quad \text{and} \quad I_{\text{bus}} = Y_{\text{bus}} \, V_{\text{bus}} \qquad (2.6)$$

Due to the generators' impedance, this admittance matrix is invertible in contrast to the power flow admittance matrix (g = generator, S = Series, P = Parallel).

$$Y_{\text{bus}} = \begin{bmatrix} y_{11} & \cdots & y_{1n} \\ \vdots & \ddots & \vdots \\ y_{n1} & \cdots & y_{nn} \end{bmatrix}, \quad \begin{cases} y_{ij} = -y_{S_{ij}} \\ y_{ii} = y_{g_i} + \displaystyle\sum_{j=1 \ne i}^{n} \left(y_{S_{ij}} + y_{P_{ij}}\right) \end{cases} \qquad (2.7)$$

$$I_{\text{bus}}^{f} = I_{\text{bus}}^{th} = I_{\text{bus}} = \begin{bmatrix} 0 & \cdots & -\hat{I}_f & \cdots & 0 \end{bmatrix}^T = \begin{bmatrix} 0 & \cdots & -\hat{I}_k^{(f)} & \cdots & 0 \end{bmatrix}^T \qquad (2.8)$$

$$V_{\text{bus}}^{th} = Z_{\text{bus}} I_{\text{bus}}; \; Z_{\text{bus}} = Y_{\text{bus}}^{-1} \Rightarrow V_{\text{bus}}^{(f)} = V_{\text{bus}}^{(o)} + Z_{\text{bus}} I_{\text{bus}} \qquad (2.9)$$

$$\begin{bmatrix} \hat{V}_1^{(f)} \\ \vdots \\ \hat{V}_k^{(f)} \\ \vdots \\ \hat{V}_n^{(f)} \end{bmatrix} = \begin{bmatrix} \hat{V}_1^{(o)} \\ \vdots \\ \hat{V}_k^{(o)} \\ \vdots \\ \hat{V}_n^{(o)} \end{bmatrix} + \begin{bmatrix} Z_{11} & \cdots & Z_{1k} & \cdots & Z_{1n} \\ \vdots & & \vdots & & \vdots \\ Z_{k1} & \cdots & Z_{kk} & \cdots & Z_{kn} \\ \vdots & & \vdots & \cdots & \vdots \\ Z_{n1} & \cdots & Z_{nk} & \cdots & Z_{nn} \end{bmatrix} \begin{bmatrix} 0 \\ \vdots \\ -\hat{I}_k^{(f)} \\ \vdots \\ 0 \end{bmatrix} \Rightarrow \qquad (2.10)$$

$$\begin{cases} \hat{V}_1^{(f)} = \hat{V}_1^{(o)} - Z_{1k}\,\hat{I}_k^{(f)} \\ \vdots \\ \hat{V}_k^{(f)} = \hat{V}_k^{(o)} - Z_{kk}\,\hat{I}_k^{(f)} \\ \vdots \\ \hat{V}_n^{(f)} = \hat{V}_n^{(o)} - Z_{nk}\,\hat{I}_k^{(f)} \end{cases}$$

Short-circuit current:

$$\Rightarrow \hat{V}_k^{(f)} = Z_f\,\hat{I}_k^{(f)} = \hat{V}_k^{(o)} - Z_{kk}\,\hat{I}_k^{(f)} \Rightarrow \hat{I}_k^{(f)} = \frac{\hat{V}_k^{(o)}}{Z_f + Z_{kk}} \qquad (2.11)$$

Post-fault bus voltages:

$$\hat{V}_i^{(f)} = \hat{V}_i^{(o)} - Z_{ik}\,\hat{I}_k^{(f)} = \hat{V}_i^{(o)} - Z_{ik}\frac{\hat{V}_k^{(o)}}{Z_f + Z_{kk}} \qquad (2.12)$$

Post-fault line currents:

$$\hat{I}_{ij}^{(f)} = \frac{\hat{V}_i^{(f)} - \hat{V}_j^{(f)}}{Z_{S_{ij}}} \qquad (2.13)$$

Note:
    $Z_{S_{ij}}$ is the series impedance between buses (i) and (j).
    Z_{ij} is the (i, j) entry of the impedance matrix.

$$Z_{S_{ij}} \neq Z_{ij} \qquad (2.14)$$

2.6   IMPEDANCE MATRIX CALCULATION

2.6.1   REVERSAL METHOD

Matrix inversion is computationally difficult.

$$Z_{\text{bus}} = Y_{\text{bus}}^{-1} \qquad (2.15)$$

2.6.2 DIRECT METHOD

By adding an impedance Z_b between the two buses, which are in one of the following states, we wish to obtain Z_{bus}^N (the new impedance matrix), assuming that Z_{bus}^O (the old impedance matrix) is known.

1- First rule: new bus and ground
2- Second rule: new and old bus
3- Third rule: old bus and ground
4- Fourth rule: old and old bus

We have:

$$V_{bus}^O = Z_{bus}^O I_{bus}^O; \quad V_{bus}^N = Z_{bus}^N I_{bus}^N \tag{2.16}$$

Keep in mind that in the matrix, Z_{0i} or Z_{oi} is the ith column and Z_{i0} or Z_{io} is the ith row.

$$Z = \begin{bmatrix} Z_{11} & \cdots & Z_{1n} \\ \vdots & & \vdots \\ Z_{n1} & \cdots & Z_{nn} \end{bmatrix} = \begin{bmatrix} Z_{1o} \\ \vdots \\ Z_{no} \end{bmatrix} = \begin{bmatrix} Z_{o1} & Z_{o2} & \cdots & Z_{on} \end{bmatrix} \tag{2.17}$$

2.6.2.1 First Rule: New Bus and Ground (Earth)

The new bus, designated $n+1$, has an impedance of Z_b and is connected to the ground. We have:

$$V_{bus}^O = Z_{bus}^O I_{bus}^O$$

$$\hat{V}_{n+1} = Z_b \hat{I}_{n+1} \Rightarrow \begin{bmatrix} \hat{V}_1 \\ \vdots \\ \hat{V}_n \\ - \\ \hat{V}_{n+1} \end{bmatrix} = \begin{bmatrix} & & & | & 0 \\ & Z_{bus}^O & & | & \vdots \\ & & & | & 0 \\ - & - & - & | & - \\ 0 & \cdots & 0 & | & Z_b \end{bmatrix} \begin{bmatrix} \hat{I}_1 \\ \vdots \\ \hat{I}_n \\ - \\ \hat{I}_{n+1} \end{bmatrix} \Rightarrow$$

$$\Rightarrow Z_{bus}^N = \begin{bmatrix} & & & | & 0 \\ & Z_{bus}^O & & | & \vdots \\ & & & | & 0 \\ - & - & - & | & - \\ 0 & \cdots & 0 & | & Z_b \end{bmatrix} \text{ or: } Z_{bus}^N = \begin{bmatrix} Z_{bus}^O & 0 \\ 0 & Z_b \end{bmatrix} \tag{2.18}$$

2.6.2.2 Second Rule: New and Old Bus

The old kth bus and the new bus, designated $n+1$, are connected via the impedance Z_b. We have:

$$\hat{V}_{n+1} = Z_b \, \hat{I}_{n+1} + \hat{V}_k \tag{2.19}$$

$$V_{\text{bus}}^O = Z_{\text{bus}}^O I_{\text{bus}}^O \Rightarrow \begin{bmatrix} \hat{V}_1 \\ \vdots \\ \hat{V}_k \\ \vdots \\ \hat{V}_n \end{bmatrix} = Z_{\text{bus}}^O \begin{bmatrix} \hat{I}_1 \\ \vdots \\ \hat{I}_k + \hat{I}_{n+1} \\ \vdots \\ \hat{I}_n \end{bmatrix} \tag{2.20}$$

The ith row:

$$\hat{V}_i = Z_{i1} \, \hat{I}_1 + \cdots + Z_{ik} \left(\hat{I}_k + \hat{I}_{n+1} \right) + \cdots + Z_{in} \, \hat{I}_n \tag{2.21}$$

The kth row:

$$\hat{V}_k = Z_{k1} \, \hat{I}_1 + \cdots + Z_{kk} \left(\hat{I}_k + \hat{I}_{n+1} \right) + \cdots + Z_{kn} \, \hat{I}_n \tag{2.22}$$

From equations 2.19 and 2.22:

$$\Rightarrow \hat{V}_{n+1} = Z_{k1} \, \hat{I}_1 + \cdots + Z_{kk}\hat{I}_k + \cdots + Z_{kn} \, \hat{I}_n + \left(Z_b + Z_{kk} \right)\hat{I}_{n+1} \tag{2.23}$$

And finally, we have:

$$\Rightarrow Z_{\text{bus}}^N = \begin{bmatrix} Z_{\text{bus}}^O & Z_{ok}^O \\ Z_{ko}^O & Z_b + Z_{kk}^O \end{bmatrix} \tag{2.24}$$

2.6.2.3 Third Rule: Old Bus and Ground (Earth)

The old kth bus has an impedance of Z_b and is connected to the ground. The following results apply if $\hat{I}_K$ is the bus current in the new network and $\hat{I}_K'$ is the bus current in the old network:

$$\text{KCL}: \hat{I}_k = \frac{\hat{V}_k}{Z_b} + \hat{I}_k' \tag{2.25}$$

$$V_{\text{bus}}^O = Z_{\text{bus}}^O I_{\text{bus}}^O \Rightarrow V = \begin{bmatrix} \hat{V}_1 \\ \vdots \\ \hat{V}_k \\ \vdots \\ \hat{V}_n \end{bmatrix} = \begin{bmatrix} Z_{\text{bus}}^O \end{bmatrix} \begin{bmatrix} \hat{I}_1 \\ \vdots \\ \hat{I}'_k \\ \vdots \\ \hat{I}_n \end{bmatrix} \Rightarrow \tag{2.26}$$

From equation 2.25:

$$V = \begin{bmatrix} Z_{\text{bus}}^O \end{bmatrix} \begin{bmatrix} \hat{I}_1 \\ \vdots \\ \hat{I}_k - \dfrac{\hat{V}_k}{Z_b} \\ \vdots \\ \hat{I}_n \end{bmatrix} = \begin{bmatrix} Z_{\text{bus}}^O \end{bmatrix} I + \begin{bmatrix} Z_{\text{bus}}^O \end{bmatrix} \begin{bmatrix} 0 \\ \vdots \\ -\dfrac{\hat{V}_k}{Z_b} \\ \vdots \\ 0 \end{bmatrix} \Rightarrow \tag{2.27}$$

$$V = \begin{bmatrix} Z_{\text{bus}}^O \end{bmatrix} I + Z_{ok}^O \left(-\dfrac{\hat{V}_k}{Z_b} \right) \tag{2.28}$$

The *kth* row:

$$\Rightarrow \hat{V}_k = Z_{ko}^O I + Z_{kk}^O \left(-\dfrac{\hat{V}_k}{Z_b} \right) \Rightarrow \hat{V}_k \left(Z_b + Z_{kk}^O \right) = Z_b Z_{ko}^O I \Rightarrow$$

$$\Rightarrow \hat{V}_k = \dfrac{Z_b}{Z_b + Z_{kk}^O} Z_{ko}^O I \tag{2.29}$$

From equations 2.28 and 2.29:

$$\Rightarrow V = \begin{bmatrix} Z_{\text{bus}}^O \end{bmatrix} I - Z_{ok}^O \left(\dfrac{1}{Z_b + Z_{kk}^O} Z_{ko}^O I \right) = \begin{bmatrix} Z_{\text{bus}}^O - \dfrac{Z_{ok}^O Z_{ko}^O}{Z_b + Z_{kk}^O} \end{bmatrix} I \tag{2.30}$$

And finally, we have:

$$\Rightarrow Z_{\text{bus}}^N = \begin{bmatrix} Z_{\text{bus}}^O - \dfrac{Z_{ok}^O Z_{ko}^O}{Z_b + Z_{kk}^O} \end{bmatrix}, \text{ or } Z_{ij}^N = Z_{ij}^O - \dfrac{Z_{ik}^O Z_{kj}^O}{Z_b + Z_{kk}^O} \tag{2.31}$$

2.6.2.4   Fourth Rule: Old and Old Bus

The two buses, k and j, that make the old buses are connected by impedance Z_b. Similar to Section 2.6.2.3, the following relationship can be demonstrated as an exercise using KVL and KCL.

$$Z_{bus}^N = Z_{bus}^O + \left[\frac{\left(Z_{ok}^O - Z_{oj}^O\right)\left(Z_{jo}^O - Z_{ko}^O\right)}{Z_b + Z_{kk}^O + Z_{jj}^O - 2Z_{kj}^O} \right] \tag{2.32}$$

$$Z_{hP}^N = Z_{hP}^O + \frac{\left(Z_{hk}^O - Z_{hj}^O\right)\left(Z_{jp}^O - Z_{kp}^O\right)}{Z_b + Z_{kk}^O + Z_{jj}^O - 2Z_{kj}^O} \tag{2.33}$$

2.7   TWO SIMPLE ALGORITHMS TO CALCULATE Z_{BUS}

2.7.1   APPROXIMATE ALGORITHM

1. All buses and ground should be added to the matrix using the first rule.
   1.1. Connect the bus to the ground using a large impedance (such as a $j1000$ pu) if it is not already connected.
2. Using the fourth (2.6.2.4) rule, add all of the remaining branches to the matrix mentioned above.

2.7.2   ACCURATE ALGORITHM

1. All buses and ground should be added to the matrix using the first rule.
   1.1. Connect the bus to the ground using an optional impedance (such as a $j1$pu) if it is not already connected.
2. Using the fourth (2.6.2.4) rule, add all of the remaining branches to the matrix mentioned above.
3. Eliminate every bus in step 1.1 using impedance $-j1$ and the third rule.

Part Two: Answer-Question

2.8   TWO-CHOICE QUESTIONS (YES/NO)

1. Is the short circuit of three phases together considered a symmetrical short circuit?
2. Does connecting two networks make both of them more susceptible to short circuits?
3. Is the network impedance matrix the inverse of the admittance matrix in load flow?
4. Are the diagonal elements of an impedance matrix equal to their Thévenin impedances at each bus?

5. Is the transformer connection type ineffective in a symmetrical short circuit?
6. Are the matrix dimensions added in the third law of direct impedance calculation?
7. Can the added impedance be eliminated from the direct impedance calculation method?
8. Should the first law always be applied when calculating impedance directly?
9. Does short-circuit current have an impact on short-circuit capacity?
10. Is it true that the inverse of the admittance matrix in a real network is more challenging than the direct method?
11. Is it possible to compute the short-circuit current using the typical load flow?
12. The series impedance between bus (i) and bus (j) is the same as the (i, j) entry of an impedance matrix?
13. Can the first and fourth laws alone be applied in the direct calculation of the impedance method rather than the entire set of four laws?
14. Should the dimensions of the impedance matrix be added in case of a new bus addition?
15. Is it necessary to calculate the impedance matrix to determine the post-fault voltage?

2.9 ANSWERS TO TWO-CHOICE QUESTIONS

1,3,6,12	No
Other	Yes

2.10 DESCRIPTIVE QUESTIONS OF THREE-PHASE SYMMETRICAL SHORT CIRCUIT

2.1 A line with impedance ($j0.1$) is added between buses 2 and 3 in a network with the following impedance matrix. What is Z_{23}^{New}?
Difficulty level o Easy •Normal o Hard

$$Z_{bus} = j \begin{bmatrix} 0.1 & 0.2 & 0.3 \\ 0.2 & 0.3 & 0.2 \\ 0.3 & 0.2 & 0.4 \end{bmatrix}$$

2.2 Find Z_{23}^{New} in question 2.1 if line 23 with impedance ($j0.1$) is removed rather than added.
Difficulty level o Easy •Normal o Hard
2.3 Find SCC for Bus 3 (Figure 2.4).
Difficulty level o Easy •Normal o Hard

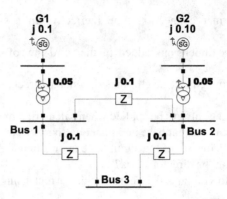

FIGURE 2.4   Question network 2.3.

2.4   When a transmission line adds a new bus to a network, how will the new SCC alter under the following two conditions?
A.   The new bus has a generator.
B.   The new bus is the load bus.
Difficulty level o Easy •Normal o Hard

2.5   (See Figure 2.5). If the CBs are open, we have $SCC_1 = 8$ and $SCC_2 = 5$. What are SCC_1 and SCC_3 if CBs closed?
Difficulty level o Easy •Normal o Hard

2.6   (See Figure 2.6) and $Z_f = 0$. If there is a short circuit in bus 3, we have: $\hat{I}_3^f = -j10$, $\hat{V}_1^f = 0.8\angle 0$, $\hat{V}_2^f = 0.5\angle 0$. What is $\hat{V}_2^f$ if there is a short circuit in bus 1.
Difficulty level o Easy o Normal • Hard

2.7   See Figure 2.1. Find Z_{bus} with matrix inversion method. $(Z_{bus} = (Y_{bus})^{-1})$
Difficulty level • Easy o Normal o Hard

2.8   Repeat question 2.7 with "Direct method" (2.6.2).
Difficulty level o Easy o Normal • Hard

2.9   Determine bus 2's post-fault voltage if bus 1 of the following network experiences a symmetrical short circuit $(Z_f = j0.1)$.
Difficulty level o Easy • Normal o Hard

$$
Z_{bus} = j \begin{bmatrix} 0.10 & 0.11 & 0.10 \\ 0.11 & 0.20 & 0.11 \\ 0.10 & 0.11 & 0.10 \end{bmatrix}
$$

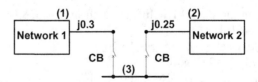

FIGURE 2.5   Question network 2.5.

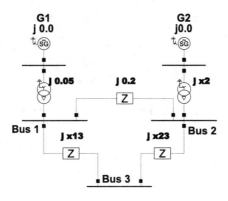

FIGURE 2.6   Question network 2.6.

2.10 With the following three matrices, construct Y_{bus} as a matrix?
Difficulty level o Easy •Normal o Hard

$$
Y_S = \begin{bmatrix} 0 & y_{S_{12}} & \cdots & y_{S_{1n}} \\ y_{S_{21}} & \ddots & & \\ \vdots & & & 0 \end{bmatrix}_{n \times n} ; \quad Y_P = \begin{bmatrix} 0 & y_{P_{12}} & \cdots & y_{P_{1n}} \\ y_{P_{21}} & \ddots & & \\ \vdots & & & 0 \end{bmatrix}_{n \times n}
$$

$$
Y_g = \begin{bmatrix} y_{g_1} & 0 & 0 \\ 0 & \ddots & 0 \\ 0 & 0 & y_{g_m} \end{bmatrix}_{m \times m} \Rightarrow Y_{bus} = ?
$$

2.11   DESCRIPTIVE ANSWERS OF THREE-PHASE SYMMETRICAL SHORT CIRCUIT

2.1 From equation 2.33: $(h = k = 2, p = j = 3)$, $Z_b = j0.1$

$$
Z_{hP}^N = Z_{hP}^O + \frac{\left(Z_{hk}^O - Z_{hj}^O\right)\left(Z_{jp}^O - Z_{kp}^O\right)}{Z_b + Z_{kk}^O + Z_{jj}^O - 2Z_{kj}^O} \Rightarrow Z_{23}^N = Z_{23}^O + \frac{\left(Z_{22}^O - Z_{23}^O\right)\left(Z_{33}^O - Z_{23}^O\right)}{Z_b + Z_{22}^O + Z_{33}^O - 2Z_{23}^O}
$$

We have:

$$
Z_{bus} = j \begin{bmatrix} 0.1 & 0.2 & 0.3 \\ 0.2 & 0.3 & 0.2 \\ 0.3 & 0.2 & 0.4 \end{bmatrix} \Rightarrow
$$

$$
Z_{23}^N = j\left(0.2 + \frac{(0.3-0.2)(0.4-0.2)}{0.1+0.3+0.4-2\times0.2} \right) = j0.25
$$

2.2   It is sufficient to parallel an impedance with a negative sign in order to remove or disconnect it.

$$(-Z) \,\|\, (Z) = \frac{(-Z)(Z)}{(-Z)+(Z)} \to \infty$$

Now, from equation 2.33: ($h = k = 2, p = j = 3$) and we have: $Z_b = (-j0.1)$:

$$Z_{hP}^N = Z_{hP}^O + \frac{\left(Z_{hk}^O - Z_{hj}^O\right)\left(Z_{jp}^O - Z_{kp}^O\right)}{Z_b + Z_{kk}^O + Z_{jj}^O - 2Z_{kj}^O} \Rightarrow Z_{23}^N = Z_{23}^O + \frac{\left(Z_{22}^O - Z_{23}^O\right)\left(Z_{33}^O - Z_{23}^O\right)}{Z_b + Z_{22}^O + Z_{33}^O - 2Z_{23}^O}$$

We have:

$$Z_{\text{bus}} = j \begin{bmatrix} 0.1 & 0.2 & 0.3 \\ 0.2 & 0.3 & 0.2 \\ 0.3 & 0.2 & 0.4 \end{bmatrix} \Rightarrow$$

$$Z_{23}^N = j\left(0.2 + \frac{(0.3-0.2)(0.4-0.2)}{-0.1+0.3+0.4-(2)(0.2)}\right) = j0.3$$

2.3   With Δ-Y conversions, we have:

$$Z_Y = \frac{(j0.1)(j0.1)}{j0.3} = j0.0333$$

The impedance seen from bus 3 is equal to:

$$Z_{33} = j0.0333 + \frac{j0.0333 + j0.15}{2} = j0.125$$

$$\Rightarrow \text{SCC}_3 = \frac{1}{|Z_{33}|} = 8 \text{ pu}$$

2.4

A.   From Section 2.6.2.3, we have:

$$Z_{ij}^N = Z_{ij}^O - \frac{Z_{ik}^O \, Z_{kj}^O}{Z_b + Z_{kk}^O} \Rightarrow X_{ii}^N = X_{ii}^O - \frac{X_{ik}^O \, X_{ki}^O}{X_b + X_{kk}^O}$$

$$\left|\frac{X_{ik}^O \, X_{ki}^O}{X_b + X_{kk}^O}\right| > 0 \Rightarrow X_{ii}^N < X_{ii}^O \to \frac{1}{X_{ii}^N} > \frac{1}{X_{ii}^O} \Rightarrow \text{SCC}^N > \text{SCC}^O$$

B. $\text{SCC}^N = \text{SCC}^O$ (Why?)

2.5 From equation 2.3 with $Z_f=0$, If the CBs are open, the Thévenin impedance is:

$$X_{11} = \frac{1}{\text{SCC}_1} = \frac{1}{8} = 0.125, \quad X_{22} = \frac{1}{\text{SCC}_2} = \frac{1}{5} = 0.2$$

If CBs closed:

$$\text{SCC}_1 = \frac{1}{X_{11}} = \frac{1}{(0.125) \| (0.3 + 0.25 + 0.2)} = 9.33$$

$$\text{SCC}_3 = \frac{1}{X_{33}} = \frac{1}{(0.125 + 0.3) \| (0.25 + 0.2)} = 4.58$$

2.6 With KVL and KCL, we have:

The short circuit is in bus 3, we have:

$$\hat{I}_{12}^f = \frac{0.8 - 0.5}{j0.2} = -j1.5, \quad \hat{I}_{g1}^f = \frac{1 - 0.8}{j0.05} = -j4 \Rightarrow \hat{I}_{13}^f = -j(4 - 1.5) = -j2.5$$

$$\Rightarrow \hat{I}_{23}^f = \hat{I}_{3}^f - \hat{I}_{13}^f = -j(10 - 2.5) = -j7.5$$

$$\Rightarrow \hat{I}_{g2}^f = \hat{I}_{23}^f - \hat{I}_{12}^f = -j(7.5 - 1.5) = -j6$$

Then:

$$jx_2 = \frac{1 - \hat{V}_2^f}{\hat{I}_{g2}} = \frac{1 - 0.5}{-j6} = j0.0833, \quad jx_{13} = \frac{\hat{V}_1^f}{\hat{I}_{13}} = \frac{0.8}{-j2.5} = j0.32$$

$$jx_{23} = \frac{\hat{V}_2^f}{\hat{I}_{23}} = \frac{0.5}{-j7.5} = j0.0667$$

If there is a short circuit in bus 1, we have:

$$\hat{I}_1^f = \frac{1}{jX_{11}} = -j\frac{1}{(x_1) \| \left[((x_{23} + x_{13}) \| x_{12}) + (x_2) \right]}$$

$$\Rightarrow \hat{I}_1^f = -j\frac{1}{(0.05) \| \left[((0.0667 + 0.32) \| 0.2) + (0.0833) \right]} =$$

$$\Rightarrow \hat{I}_1^f = -j\frac{1}{(0.05) \| (0.215)} = -j24.65$$

KCL: (How?)

$$\hat{I}_{g2}^f = \frac{j0.05}{j0.05 + j0.215}(-j24.65) = -j4.651$$

Then:

$$\Rightarrow \hat{V}_2^f = 1 - jx_2\, \hat{I}_{g2}^f = 1 - (j0.0833)(-j4.651) = 0.613\angle 0$$

2.7 (See Figure 2.1 or 2.2). The impedance network must be transformed into an admittance network, similar to Figure 2.7.
From equation 2.7:

$$Y_{bus} = -j\begin{bmatrix} 3.333 + 2.5 + 1.25 & -2.5 & -1.25 \\ -2.5 & 3.333 + 2.5 + 1.25 & -1.25 \\ -1.25 & -1.25 & 1.25 + 1.25 \end{bmatrix}$$

$$\Rightarrow Y_{bus} = -j\begin{bmatrix} 7.083 & -2.5 & -1.25 \\ -2.5 & 7.083 & -1.25 \\ -1.25 & -1.25 & 2.5 \end{bmatrix}$$

$$Z_{bus} = Y_{bus}^{-1} = j\begin{bmatrix} 0.2022 & 0.0978 & 0.1500 \\ 0.0978 & 0.2022 & 0.1500 \\ 0.1500 & 0.1500 & 0.5500 \end{bmatrix}$$

Note that there is an equal Z_{33} from the Thévenin method (Figure 2.3) and impedance matrix. ($Z_{33} = j0.55$)

2.8 (See Figure 2.7). Following the guidelines in 2.6.2, five impedances, Z_1–Z_5, are added to the impedance matrix.
Z_1: (2.6.2.1 new bus and ground). From equation 2.18:

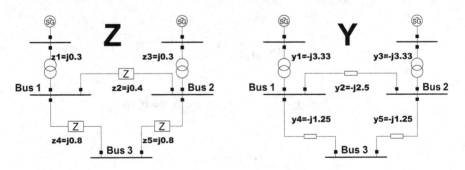

FIGURE 2.7 Answer network 2.7.

$$Z_{bus}^N = \begin{bmatrix} Z_{bus}^O & 0 \\ 0 & Z_b \end{bmatrix}, \text{ and, } Z_b = Z_1 = j0.3, \ Z_{bus}^O = \varnothing \Rightarrow$$

$$Z_{bus}^N = j0.3$$

Z_2: (2.6.2.2 new and old bus). From equation 2.24:

$$Z_{bus}^N = \begin{bmatrix} Z_{bus}^O & Z_{ok}^O \\ Z_{ko}^O & Z_b + Z_{kk}^O \end{bmatrix}, k = 1, \ Z_{bus}^O = j0.3, \ Z_b = j0.4 \Rightarrow,$$

$$Z_{bus}^N = j \begin{bmatrix} 0.3 & 0.3 \\ 0.3 & 0.3+0.4 \end{bmatrix} = j \begin{bmatrix} 0.3 & 0.3 \\ 0.3 & 0.7 \end{bmatrix}$$

Z_3: (2.6.2.3 old bus and ground). From equation 2.31:

$$Z_{bus}^N = \begin{bmatrix} Z_{bus}^O - \dfrac{Z_{ok}^O Z_{ko}^O}{Z_b + Z_{kk}^O} \end{bmatrix}, k = 2, \ Z_{bus}^O = j \begin{bmatrix} 0.3 & 0.3 \\ 0.3 & 0.7 \end{bmatrix}, \ Z_b = j0.3 \Rightarrow,$$

$$Z_{bus}^N = j \left(\begin{bmatrix} 0.3 & 0.3 \\ 0.3 & 0.7 \end{bmatrix} - \dfrac{1}{0.3+0.7} \begin{bmatrix} 0.3 \\ 0.7 \end{bmatrix} \begin{bmatrix} 0.3 & 0.7 \end{bmatrix} \right) \Rightarrow$$

$$Z_{bus}^N = j \left(\begin{bmatrix} 0.3 & 0.3 \\ 0.3 & 0.7 \end{bmatrix} - \dfrac{1}{1.0} \begin{bmatrix} 0.09 & 0.21 \\ 0.21 & 0.49 \end{bmatrix} \right) = j \begin{bmatrix} 0.21 & 0.09 \\ 0.09 & 0.21 \end{bmatrix}$$

Z_4: (2.6.2.2 new and old bus). From equation 2.24:

$$Z_{bus}^N = \begin{bmatrix} Z_{bus}^O & Z_{ok}^O \\ Z_{ko}^O & Z_b + Z_{kk}^O \end{bmatrix}, k = 1, \ Z_{bus}^O = j \begin{bmatrix} 0.21 & 0.09 \\ 0.09 & 0.21 \end{bmatrix}, \ Z_b = j0.8 \Rightarrow$$

$$Z_{bus}^N = j \begin{bmatrix} 0.21 & 0.09 & 0.21 \\ 0.09 & 0.21 & 0.09 \\ 0.21 & 0.09 & 0.21+0.8 \end{bmatrix}$$

Z_5: (2.6.2.4 old and old bus). From equation 2.32:

$$Z_{bus}^N = Z_{bus}^O + \begin{bmatrix} \dfrac{(Z_{ok}^O - Z_{oj}^O)(Z_{jo}^O - Z_{ko}^O)}{Z_b + Z_{kk}^O + Z_{jj}^O - 2Z_{kj}^O} \end{bmatrix}, k = 2, \ j = 3, \ Z_b = j0.8 \Rightarrow,$$

$$Z_{bus}^O = j \begin{bmatrix} 0.21 & 0.09 & 0.21 \\ 0.09 & 0.21 & 0.09 \\ 0.21 & 0.09 & 1.01 \end{bmatrix} \Rightarrow$$

$$\Rightarrow Z_b + Z_{kk}^O + Z_{jj}^O - 2 \times Z_{kj}^O = j(0.8 + 0.21 + 1.01 - 2 \times 0.09) = j1.84$$

$$\Rightarrow \left(Z_{ok}^O - Z_{oj}^O \right) = j \begin{bmatrix} 0.09 - 0.21 \\ 0.21 - 0.09 \\ 0.09 - 1.01 \end{bmatrix} = j \begin{bmatrix} -0.12 \\ 0.12 \\ -0.92 \end{bmatrix}$$

$$\Rightarrow \left(Z_{jo}^O - Z_{ko}^O \right) = j \begin{bmatrix} 0.12 & -0.12 & 0.92 \end{bmatrix}$$

$$\Rightarrow \left(Z_{ok}^O - Z_{oj}^O \right)\left(Z_{jo}^O - Z_{ko}^O \right) = j^2 \begin{bmatrix} -0.12 \\ 0.12 \\ -0.92 \end{bmatrix} \begin{bmatrix} 0.12 & -0.12 & 0.92 \end{bmatrix} \Rightarrow$$

$$\Rightarrow \left(Z_{ok}^O - Z_{oj}^O \right)\left(Z_{jo}^O - Z_{ko}^O \right) = \begin{bmatrix} 0.0144 & -0.0144 & 0.1104 \\ -0.0144 & 0.0144 & -0.1104 \\ 0.1104 & -0.1104 & 0.8464 \end{bmatrix}$$

$$Z_{bus}^N = j \begin{bmatrix} 0.21 & 0.09 & 0.21 \\ 0.09 & 0.21 & 0.09 \\ 0.21 & 0.09 & 1.01 \end{bmatrix} + \frac{1}{j1.84} \begin{bmatrix} 0.0144 & -0.0144 & 0.1104 \\ -0.0144 & 0.0144 & -0.1104 \\ 0.1104 & -0.1104 & 0.8464 \end{bmatrix}$$

$$\Rightarrow Z_{bus} = j \begin{bmatrix} 0.2022 & 0.0978 & 0.1500 \\ 0.0978 & 0.2022 & 0.1500 \\ 0.1500 & 0.1500 & 0.5500 \end{bmatrix}$$

The response to the previous question is the same as this one.

2.9   From equation 2.11:

$$k = 1, \Rightarrow \hat{I}_1^{(f)} = \frac{\hat{V}_1^{(o)}}{Z_f + Z_{11}} = \frac{1}{j0.1 + j0.1} = -j5$$

From equation 2.12:

$$i = 2, \quad \hat{V}_2^{(f)} = \hat{V}_2^{(o)} - Z_{21} I_1^{(f)} = 1 - j(0.11)(-j5) = 1 - 0.55 = 0.45$$

3 Three-Phase Unsymmetrical Short Circuit

Part One: Lesson Summary

3.1 INTRODUCTION

This chapter reviews the calculations of a three-phase unsymmetrical short-circuit currents. First, its symmetrical components and sequence impedances (+, −, 0) are presented. Next, the symmetrical components model for star-connected load, symmetrical transmission line, symmetrical three-phase transformer, and loaded generator is given. In the end, the equivalent circuit of four types of unsymmetrical short circuits is specified, which are: single phase-to-ground, phase-to-phase (two-phase), two phase-to-ground, and phase-to-phase, other single phase-to-ground. It has been demonstrated that the most significant short circuit for generators is a single-phase short circuit. This kind of short circuit is also the most frequent one. Lastly, to perform unsymmetrical short-circuit calculations, symmetrical short-circuit calculations are required.

3.2 SYMMETRICAL COMPONENTS

Assume that a set of three-phase voltages designated is given. In accordance with Fortescue, these phase voltages can be decomposed into three sets of sequence components as follows:

- The positive sequence components are typically denoted by the number 1 and follow the same sequence as +ABC (or balanced +) have the same magnitude, phase sequence, and direction of rotation as the original three-phase system.
- The negative sequence components are typically denoted by the number 2. They follow the same sequence as -ABC (or balanced −).
- The number 0 is typically used to present zero sequence components, which are unbalanced components or zero sequences, and there's no phase shift or rotation involved.

  Note: the term "rotation" is metaphorical when describing electrical sequences, it is used to illustrate the relative behavior of the symmetrical components in a three-phase system rather than literal spinning or movement.

DOI: 10.1201/9781003506751-3

$$\begin{cases} \hat{I}_a = \hat{I}_a^+ + \hat{I}_a^- + \hat{I}_a^0 \\ \hat{I}_b = \hat{I}_b^+ + \hat{I}_b^- + \hat{I}_b^0 \\ \hat{I}_c = \hat{I}_c^+ + \hat{I}_c^- + \hat{I}_c^0 \end{cases} \qquad (3.1)$$

α is a complex number with a unit magnitude and a 120° phase angle involving ($\alpha \triangleq 1\angle 120°$). Common identities are involving:

$$\begin{aligned} \alpha^4 &= \alpha = 1\angle 120° \\ \alpha^2 &= 1\angle 240° = \alpha* \\ \alpha^3 &= 1\angle 0° \\ 1 + \alpha + \alpha^2 &= 0 \\ 1 - \alpha &= \sqrt{3}\angle - 30° \\ 1 - \alpha^2 &= \sqrt{3}\angle + 30° \\ \alpha^2 - \alpha &= \sqrt{3}\angle 270° \\ j\alpha &= 1\angle 210° \\ 1 + \alpha &= -\alpha^2 = 1\angle 60° \\ 1 + \alpha^2 &= -\alpha = 1\angle - 60° \\ \alpha + \alpha^2 &= -1 = 1\angle 180° \end{aligned} \qquad (3.2)$$

From Figure 3.1, we have:

$$\begin{cases} \hat{I}_b^0 = \hat{I}_c^0 = \hat{I}_a^0 \\ \hat{I}_b^+ = \alpha^2 \hat{I}_a^+, \ \hat{I}_c^+ = \alpha \hat{I}_a^+, \Rightarrow \\ \hat{I}_b^- = \alpha \hat{I}_a^-, \ \hat{I}_c^- = \alpha^2 \hat{I}_a^- \end{cases} \quad \overbrace{\begin{bmatrix} \hat{I}_a \\ \hat{I}_b \\ \hat{I}_c \end{bmatrix}}^{I^{abc}} = \overbrace{\begin{bmatrix} 1 & 1 & 1 \\ 1 & \alpha^2 & \alpha \\ 1 & \alpha & \alpha^2 \end{bmatrix}}^{A} \times \overbrace{\begin{bmatrix} \hat{I}_a^0 \\ \hat{I}_a^+ \\ \hat{I}_a^- \end{bmatrix}}^{I_a^{0+-}} \qquad (3.3)$$

or:

$$\Rightarrow I^{abc} = A I_a^{0+-} \qquad \Rightarrow I_a^{0+-} = A^{-1} I^{abc} \qquad (3.4)$$

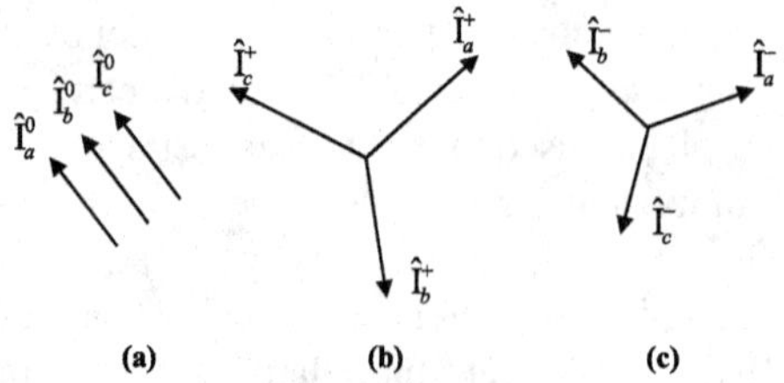

(a)   (b)   (c)

FIGURE 3.1   Resolving phase currents into three sets of sequence components.

Also, for the voltages we have:

$$V^{abc} = AV_a^{0+-}, \quad V_a^{0+-} = A^{-1}V^{abc} \tag{3.5}$$

A is defined as the symmetrical component transformation matrix.

Exercise 3.1:
Show that:

$$A^{-1} = \frac{1}{3}A^* = \frac{1}{3}\begin{bmatrix} 1 & 1 & 1 \\ 1 & \alpha & \alpha^2 \\ 1 & \alpha^2 & \alpha \end{bmatrix} \tag{3.6}$$

For the apparent powers we have:

$$S_{3ph} = \hat{V}_a\hat{I}_a^* + \hat{V}_b\hat{I}_b^* + \hat{V}_c\hat{I}_c^* \tag{3.7}$$

Power formula according to the symmetrical components:

$$S_{3ph} = \begin{bmatrix} \hat{V}_a & \hat{V}_b & \hat{V}_c \end{bmatrix}\begin{bmatrix} \hat{I}_a \\ \hat{I}_b \\ \hat{I}_c \end{bmatrix}^* = \left(V^{abc}\right)^T \left(I^{abc}\right)^* =$$

$$= \left(AV_a^{0+-}\right)^T \left(AI_a^{0+-}\right)^* = (V_a^{0+-})^T A^T A^* (I_a^{0+-})^* \tag{3.8}$$

We have:

$$\begin{cases} A^T = A \\ A^T A^* = A(3A^{-1}) = 3I_{3\times3} \end{cases} \tag{3.9}$$

$$\Rightarrow S_{3ph} = 3\left(V_a^{0+-}\right)^T \left(I_a^{0+-}\right)^* = 3\left(\hat{V}_a^0\,\hat{I}_a^{0*} + \hat{V}_a^+\,\hat{I}_a^{+*} + \hat{V}_a^-\,\hat{I}_a^{-*}\right) \tag{3.10}$$

3.3   SEQUENCE IMPEDANCES (+, −, 0)

3.3.1   STAR-CONNECTED LOAD

(Z_m) is the mutual impedance and (Z_s) is the series impedance of each phase (Figure 3.2).

$$V_a^{0+-} = Z^{0+-}I_a^{0+-}$$

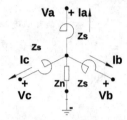

FIGURE 3.2   Star load equivalent circuit.

$$Z^{0+-} = \begin{bmatrix} Z_s + 3Z_n + 2Z_m & 0 & 0 \\ 0 & Z_s - Z_m & 0 \\ 0 & 0 & Z_s - Z_m \end{bmatrix}, \; or \; \begin{bmatrix} Z^0 \\ Z^+ \\ Z^- \end{bmatrix} = \begin{bmatrix} Z_s + 3Z_n + 2Z_m \\ Z_s - Z_m \\ Z_s - Z_m \end{bmatrix}$$

(3.11)

In the zero network, neutral impedance (Z_n) appears three times $(3Z_n)$. Neither the positive nor the negative networks exhibit neutral impedance.

3.3.2 SYMMETRIC TRANSMISSION LINE

Due to the line transposition,[1] we have:

$$Z^+_{\text{Line}} = Z^-_{\text{Line}}$$

(3.12)

Assuming an unrealistic neutral wire inside the ground, we have:

$$X^0 = X^+ + 3X_n \implies X^0 = (3 \, or \, 4)X^+$$

(3.13)

In equation 3.13, you can see the earth's effect three times. Due to this, the zero component of the line has a much higher impedance than the positive and negative components.

3.3.3 SYMMETRIC THREE-PHASE TRANSFORMER

Since the transformer is symmetrical, we have:

$$Z^+ = Z^- = Z^0$$

(3.14)

Simply focus on the transformers' zero sequence model. The zero component of the models listed below is disconnected (Figure 3.3).

The zero component of the models listed below is connected (Figure 3.4).

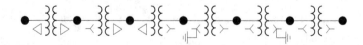

FIGURE 3.3   Disconnected models at zero components.

FIGURE 3.4 Connected models at zero components.

3.3.4 LOADED GENERATOR

$$\text{if}: E^{abc} = \begin{bmatrix} E_a \angle 0 \\ E_a \angle -120 \\ E_a \angle 120 \end{bmatrix} \Rightarrow V^{0+-} = E^{0+-} - Z^{0+-} I^{0+-}, \ E^{0+-} = \begin{bmatrix} 0 \\ E_a \\ 0 \end{bmatrix} \quad (3.15)$$

$$\begin{bmatrix} Z^0 \\ Z^+ \\ Z^- \end{bmatrix} = \begin{bmatrix} Z_s + 3Z_n \\ Z_s \\ Z_s \end{bmatrix} \quad (3.16)$$

The loaded generator's final equivalent circuit is displayed as the following (Figures 3.5 and 3.6).
Note that:

$$X_g^+ = X_d'' \ or \ X_d' \ or \ X_d, \ X_g^- = X_d'', X_g^0 = X_l \simeq 0,$$

$$\Rightarrow X_g^+ \geq X_g^- > X_g^0 \geq 0$$

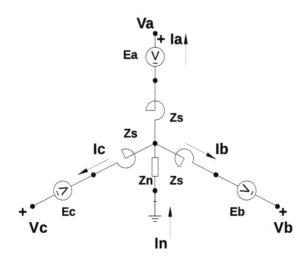

FIGURE 3.5 Loaded generator.

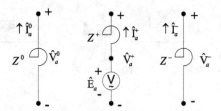

FIGURE 3.6 The loaded generator's final equivalent circuit.

3.4 SHORT CIRCUIT EQUIVALENT CIRCUIT

3.4.1 SINGLE PHASE-TO-GROUND

When a single-phase short circuit to the ground occurs, it is the most dangerous when it occurs close to the generator, and least dangerous when it occurs reasonably far from it. The occurrence likelihood of this short circuit is high (up to 95%) (Figure 3.7).

Short-circuit current is:

$$\begin{bmatrix} \hat{I}_a \\ 0 \\ 0 \end{bmatrix} = \begin{bmatrix} \hat{I}_f \\ 0 \\ 0 \end{bmatrix}; \quad \hat{V}_a = Z_f \hat{I}_f = Z_f \hat{I}_a \tag{3.17}$$

$$\Rightarrow \begin{bmatrix} \hat{I}_a^0 \\ \hat{I}_a^+ \\ \hat{I}_a^- \end{bmatrix} = \frac{1}{3} \begin{bmatrix} 1 & 1 & 1 \\ 1 & \alpha & \alpha^2 \\ 1 & \alpha^2 & \alpha \end{bmatrix} \begin{bmatrix} \hat{I}_a \\ 0 \\ 0 \end{bmatrix} \Rightarrow \hat{I}_a^0 = \hat{I}_a^+ = \hat{I}_a^- = \frac{\hat{I}_a}{3} \tag{3.18}$$

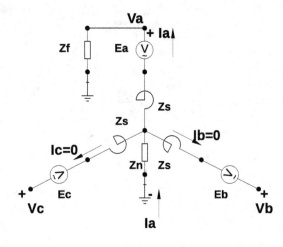

FIGURE 3.7 Real circuit of a single phase-to-ground short circuit

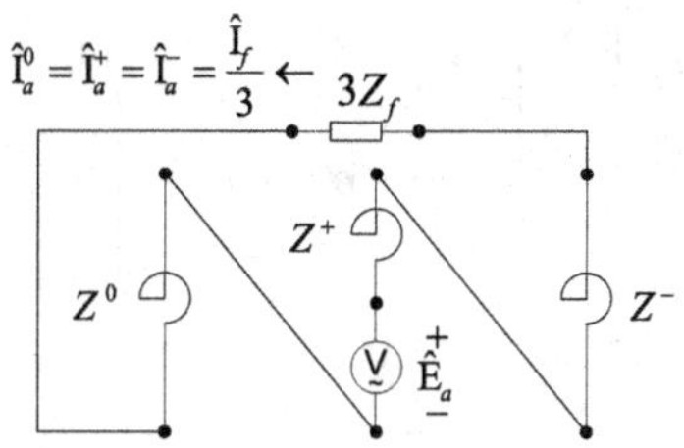

$$\hat{I}_a^0 = \hat{I}_a^+ = \hat{I}_a^- = \frac{\hat{I}_f}{3} \leftarrow \quad 3Z_f$$

$$Z^+$$

$$Z^0 \qquad Z^-$$

$$V \quad \hat{E}_a^+$$

FIGURE 3.8 Equivalent circuit of single phase-to-ground short circuit.

$$\hat{V}_a = Z_f \hat{I}_a \Rightarrow \quad \hat{V}_a^0 + \hat{V}_a^+ + \hat{V}_a^- = Z_f \hat{I}_a \quad \xrightarrow{(3.18)} \quad \hat{V}_a^0 + \hat{V}_a^+ + \hat{V}_a^- = 3Z_f \hat{I}_a^0 \quad (3.19)$$

$$\Rightarrow \hat{I}_a^0 = \hat{I}_a^+ = \hat{I}_a^- = \frac{\hat{I}_a}{3} = \frac{\hat{I}_f}{3}; \quad \begin{cases} \hat{V}_a^0 = -Z^0 \hat{I}_a^0 \\ \hat{V}_a^+ = \hat{E}_a - Z^+ \hat{I}_a^+ \\ \hat{V}_a^- = -Z^- \hat{I}_a^- \end{cases} \Rightarrow \frac{\hat{I}_f}{3} = \frac{\hat{E}_a}{Z^+ + Z^- + Z^0 + 3Z_f} \quad (3.20)$$

Equation 3.20 can be represented as Figure 3.8's equivalent circuit.

3.4.2 PHASE-TO-PHASE, TWO-PHASE

Short-circuit current is (Figure 3.9):

$$\begin{bmatrix} 0 \\ \hat{I}_f \\ -\hat{I}_f \end{bmatrix} = \begin{bmatrix} 0 \\ \hat{I}_b \\ -\hat{I}_b \end{bmatrix} \quad (3.21)$$

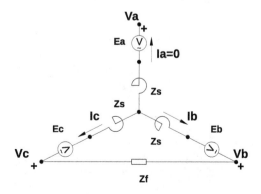

FIGURE 3.9 Real circuit of phase-to-phase, two-phase short circuit.

$$\hat{V}_b - \hat{V}_c = Z_f \hat{I}_f, \quad \begin{bmatrix} \hat{I}_a^0 \\ \hat{I}_a^+ \\ \hat{I}_a^- \end{bmatrix} = \frac{1}{3} \begin{bmatrix} 1 & 1 & 1 \\ 1 & \alpha & \alpha^2 \\ 1 & \alpha^2 & \alpha \end{bmatrix} \begin{bmatrix} 0 \\ \hat{I}_f \\ -\hat{I}_f \end{bmatrix} = \frac{1}{3} \begin{bmatrix} 0 \\ (\alpha - \alpha^2) \\ (\alpha^2 - \alpha) \end{bmatrix} \hat{I}_f \quad (3.22)$$

$$\hat{I}_a = \hat{I}_a^0 + \hat{I}_a^+ + \hat{I}_a^-; \quad \hat{I}_a^0 = 0; \quad \hat{I}_a^+ = -\hat{I}_a^- = \frac{(\alpha - \alpha^2)\hat{I}_f}{3} = \frac{j\sqrt{3}}{3}\hat{I}_f \quad (3.23)$$

$$\hat{V}_b - \hat{V}_c = Z_f \hat{I}_f \Rightarrow \left(\hat{V}_a^0 + \alpha^2 \hat{V}_a^+ + \alpha \hat{V}_a^-\right) - \left(\hat{V}_a^0 + \alpha \hat{V}_a^+ + \alpha^2 \hat{V}_a^-\right) = Z_f \hat{I}_f \quad (3.24)$$

$$\Rightarrow \left(\alpha^2 - \alpha\right)\hat{V}_a^+ + \left(\alpha - \alpha^2\right)\hat{V}_a^- = Z_f \hat{I}_f \quad \Rightarrow \quad \hat{V}_a^+ - \hat{V}_a^- = \frac{Z_f}{\alpha^2 - \alpha}\hat{I}_f \quad (3.25)$$

$$\overset{(3.23)}{\Rightarrow} \quad \hat{V}_a^+ - \hat{V}_a^- = \frac{3\hat{I}_a^+ Z_f}{\left(\alpha^2 - \alpha\right)\left(\alpha - \alpha^2\right)} \quad (3.26)$$

From equation 3.2, we have:

$$\left(\alpha^2 - \alpha\right)\left(\alpha - \alpha^2\right) = \alpha^3 - \alpha^4 - \alpha^2 + \alpha^3 = 1 - \alpha - \alpha^2 + 1 = 3 \quad (3.27)$$

$$\Rightarrow \hat{V}_a^+ - \hat{V}_a^- = Z_f \hat{I}_a^+ \quad (3.28)$$

Equations 3.28 and 3.23 can be represented as Figure 3.10's equivalent circuit.

3.4.3 TWO PHASE-TO-GROUND

Short-circuit current is (Figure 3.11):

$$\hat{I}_f = \hat{I}_b + \hat{I}_c \quad (3.29)$$

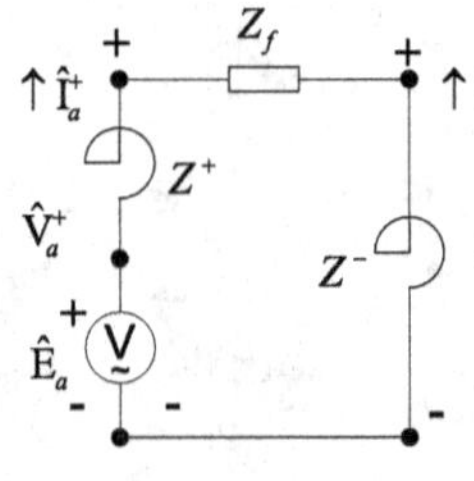

FIGURE 3.10 Equivalent circuit of phase-to-phase, two-phase short circuit.

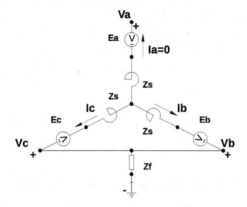

FIGURE 3.11 Real circuit of a two phase-to-ground short circuit.

$$\text{KCL}: \begin{bmatrix} 0 \\ \hat{I}_b \\ \hat{I}_f - \hat{I}_b \end{bmatrix} = \begin{bmatrix} 0 \\ \hat{I}_b \\ \hat{I}_c \end{bmatrix}, \quad \text{KVL}: \quad \hat{V}_b = \hat{V}_c = Z_f \hat{I}_f \tag{3.30}$$

$$\begin{bmatrix} \hat{I}_a^0 \\ \hat{I}_a^+ \\ \hat{I}_a^- \end{bmatrix} = \frac{1}{3} \begin{bmatrix} 1 & 1 & 1 \\ 1 & \alpha & \alpha^2 \\ 1 & \alpha^2 & \alpha \end{bmatrix} \begin{bmatrix} 0 \\ \hat{I}_b \\ \hat{I}_f - \hat{I}_b \end{bmatrix} = \frac{1}{3} \begin{bmatrix} \hat{I}_f \\ (\alpha - \alpha^2)\hat{I}_b + \alpha^2 \hat{I}_f \\ (\alpha^2 - \alpha)\hat{I}_b + \alpha \hat{I}_f \end{bmatrix} \Rightarrow \hat{I}_a^0 = \frac{\hat{I}_f}{3} \tag{3.31}$$

We have:

$$\hat{I}_a = 0 = \hat{I}_a^0 + \hat{I}_a^+ + \hat{I}_a^- \tag{3.32}$$

$$\hat{V}_b = \hat{V}_c \Rightarrow \hat{V}_a^0 + \alpha^2 \hat{V}_a^+ + \alpha \hat{V}_a^- = \hat{V}_a^0 + \alpha \hat{V}_a^+ + \alpha^2 \hat{V}_a^-$$

$$\Rightarrow (\alpha^2 - \alpha)\hat{V}_a^+ = (\alpha^2 - \alpha)\hat{V}_a^- \Rightarrow \hat{V}_a^+ = \hat{V}_a^- \tag{3.33}$$

$$\hat{V}_b = \hat{V}_c = Z_f \hat{I}_f \Rightarrow \hat{V}_a^0 + \alpha^2 \hat{V}_a^+ + \alpha \hat{V}_a^- = Z_f \hat{I}_f \Rightarrow \hat{V}_a^0 + (\alpha^2 + \alpha)\hat{V}_a^+ = Z_f \hat{I}_f \tag{3.34}$$

$$\Rightarrow \hat{V}_a^0 - \hat{V}_a^+ = Z_f \hat{I}_f \stackrel{(3.31)}{=} Z_f \left(3\hat{I}_a^0\right) \Rightarrow \hat{V}_a^0 - \hat{V}_a^+ = \left(3Z_f\right)\hat{I}_a^0 \tag{3.35}$$

Equations 3.31, 3.33, and 3.35 can be represented as Figure 3.12's equivalent circuit.

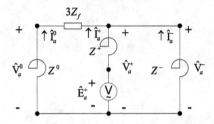

FIGURE 3.12 Equivalent circuit of two phase-to-ground short circuit.

3.4.4 PHASE-TO-PHASE, OTHER SINGLE PHASE-TO-GROUND

Let's consider the real circuit of phase-to-phase, other single phase-to-ground.
  Short-circuit current is:

$$\text{KCL}:\begin{bmatrix} \hat{I}_a \\ \hat{I}_b \\ \hat{I}_c \end{bmatrix} = \begin{bmatrix} \hat{I}_{f1} \\ \hat{I}_{f2} \\ -\hat{I}_{f2} \end{bmatrix} \tag{3.36}$$

$$\Rightarrow \begin{bmatrix} \hat{I}_a^0 \\ \hat{I}_a^+ \\ \hat{I}_a^- \end{bmatrix} = \frac{1}{3}\begin{bmatrix} 1 & 1 & 1 \\ 1 & \alpha & \alpha^2 \\ 1 & \alpha^2 & \alpha \end{bmatrix}\begin{bmatrix} \hat{I}_{f1} \\ \hat{I}_{f2} \\ -\hat{I}_{f2} \end{bmatrix} = \frac{1}{3}\begin{bmatrix} \hat{I}_{f1} \\ \hat{I}_{f1}+\left(\alpha-\alpha^2\right)\hat{I}_{f2} \\ \hat{I}_{f1}+\left(\alpha^2-\alpha\right)\hat{I}_{f2} \end{bmatrix} \tag{3.37}$$

$$\Rightarrow \hat{I}_a^+ + \hat{I}_a^- = \frac{2}{3}\hat{I}_{f1} = 2\hat{I}_a^0 \tag{3.38}$$

$$\begin{cases} \hat{V}_a = 0 = \hat{V}_a^0 + \hat{V}_a^+ + \hat{V}_a^- \\ \hat{V}_b = \hat{V}_c \quad \Rightarrow \quad \hat{V}_a^0 + \alpha^2\hat{V}_a^+ + \alpha\hat{V}_a^- = \hat{V}_a^0 + \alpha\hat{V}_a^+ + \alpha^2\hat{V}_a^- \end{cases} \Rightarrow \hat{V}_a^+ = \hat{V}_a^- \tag{3.39}$$

$$\Rightarrow \hat{V}_a^0 = -2\hat{V}_a^+ \tag{3.40}$$

Finding the equivalent circuit is the reader's responsibility. Utilize the loaded
generator's equivalent circuit as well as equations 3.38–3.40. Following the circuit's
drawing, the circuit's solution can be computed as follows.

$$\hat{I}_a^+ = \frac{\hat{E}}{jx^+ +\left(jx^- \parallel \dfrac{jx^0}{4}\right)},\quad \hat{I}_a^- = \frac{-\dfrac{jx^0}{4}}{jx^- + \dfrac{jx^0}{4}}\hat{I}_a^+,\quad 2\hat{I}_a^0 = \frac{jx^-}{jx^- + \dfrac{jx^0}{4}}\hat{I}_a^+ \tag{3.41}$$

It is assumed that: $Z_{f1} = Z_{f2} = 0$.

Exercise 3.2:
Obtain the new relations by assuming $Z_{f1} \neq Z_{f2} \neq 0$.

Part Two: Answer-Question

3.5 TWO-CHOICE QUESTIONS (YES/NO)

1. The single-phase short-circuit current is eliminated by isolating the generators and transformers.
2. The two-phase short circuit is eliminated by isolating the generators and transformers.
3. There is no effect from the transformer connections in the single-phase short-circuit current.
4. There is no effect from the transformer connections in the two-phase short circuit.
5. Always $(Z^+ > Z^-)$.
6. This formula is correct $(X_g^+ \geq X_g^- > X_g^0 \geq 0)$.
7. Calculation of single-phase current is easier than two-phase.
8. In the zero sequence network, neutral impedance (Z_n) appears three times $(3Z_n)$.
9. "Zero sequence impedance" appears six times in the YY connection grounded with neutral impedance.
10. Reducing the fault current involves cutting off the line.
11. In theoretical calculations, the fault current is larger than the reality.
12. This formula is correct $\left(S_{3ph} = \hat{V}_a^0 \hat{I}_a^{0*} + \hat{V}_a^+ \hat{I}_a^{+*} + \hat{V}_a^- \hat{I}_a^{-*} \right)$.
13. This formula is correct $\left(S_{3ph} = \hat{V}_a \hat{I}_a^* + \hat{V}_b \hat{I}_b^* + \hat{V}_c \hat{I}_c^* \right)$.
14. This formula is correct $\left(\left| \alpha - \alpha^2 / \alpha + \alpha^2 \right| = \sqrt{3} \right)$.
15. This formula is correct $\left(A^{-1} = A^* \right)$.
16. Symmetric fault analysis is easier than asymmetric fault calculations.
17. Generators are the most vulnerable power system components to a single-phase fault.
18. It is easier to mitigate single-phase faults compared to three-phase faults.
19. Typical loads significantly impact the fault current.
20. In fault studies, all nonrotating impedance loads are usually neglected.
21. Can superposition be applied in short-circuit studies for calculating fault currents?
22. Before proceeding with per-unit fault current calculations, considering the single-line diagram of the network, a positive-sequence equivalent circuit is set up on a chosen base system.

3.6   ANSWERS TO TWO-CHOICE QUESTIONS

| 2,3,5,7,12,15,19 | No |
| Other | Yes |

3.7   DESCRIPTIVE QUESTIONS OF THREE-PHASE UNSYMMETRICAL SHORT CIRCUIT

3.1   Find the symmetrical components of the following three-phase current.
Difficulty level o Easy •Normal o Hard

$$
\begin{cases}
\hat{I}_a = 1.6\angle 25^\circ \ \text{pu} \\[2mm]
\hat{I}_b = 1\angle 180^\circ \quad \text{pu} \\[2mm]
\hat{I}_c = 0.9\angle 132^\circ \ \text{pu}
\end{cases}
$$

3.2   Find the symmetrical components transformation matrix in the following form.
Difficulty level • Easy o Normal o Hard

$$
I^{abc} = A I_a^{0+-} \Rightarrow
\overbrace{\begin{bmatrix} \hat{I}_a \\ \hat{I}_b \\ \hat{I}_c \end{bmatrix}}^{I^{abc}}
=
\overbrace{\begin{bmatrix} 1 & 1 & 1 \\ 1 & \alpha^2 & \alpha \\ 1 & \alpha & \alpha^2 \end{bmatrix}}^{A}
\overbrace{\begin{bmatrix} \hat{I}_a^0 \\ \hat{I}_a^+ \\ \hat{I}_a^- \end{bmatrix}}^{I_a^{0+-}},
\quad
\overbrace{\begin{bmatrix} \hat{I}_a \\ \hat{I}_b \\ \hat{I}_c \end{bmatrix}}^{I^{abc}}
=
\overbrace{[?]}^{A^{New}}
\overbrace{\begin{bmatrix} \hat{I}_a^+ \\ \hat{I}_a^- \\ \hat{I}_a^0 \end{bmatrix}}^{I_a^{+-0}}
$$

3.3   If the current in two phases is as follows, find the symmetrical components in a delta load.

$$
\hat{I}_a = 10\angle 30, \quad \hat{I}_b = 15\angle -60
$$

Difficulty level o Easy •Normal o Hard

3.4   A balanced three-phase source is connected to phase A and ground through a 10 Ω resistance, and phases B and C are connected to a $j10$ Ω reactance. Determine the symmetrical components of the current if $V_L = 4.8$ kV.
Difficulty level o Easy •Normal o Hard

3.5   Determine and compare the power consumption of question (3.4) using various techniques.
Difficulty level o Easy •Normal o Hard

3.6   For the following network, draw the symmetrical components (+, −, 0) (Figure 3.13).
Difficulty level o Easy •Normal o Hard

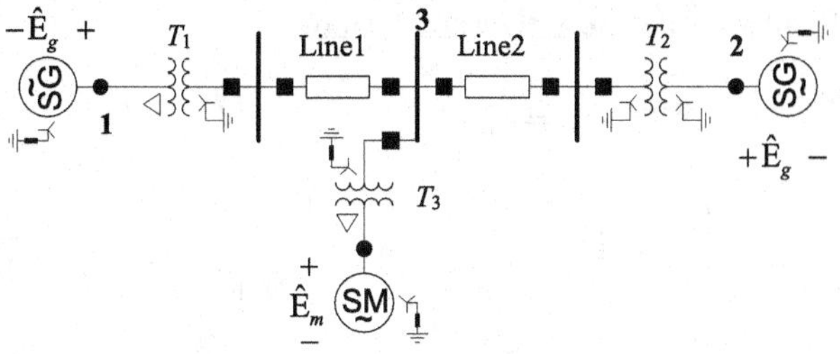

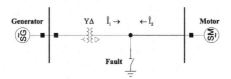

FIGURE 3.13 Question network 3.6.

3.7 Concerning the network of question **3.6**, the following details apply. Acquire the reactances in the $(+, -, 0)$ sequence from Bus 3's viewpoint.

$$G_1, G_2 : 20 \text{ MVA}, 13 \text{ kV}, X^+ = X^- = 20\%, X^0 = 8\%, X_n = 5\%$$

$$M : 30 \text{ MVA}, 13.8 \text{ kV}, X^+ = X^- = 20\%, X^0 = 8\%, X_n = 5\%$$

$$T_{Y-Y} : 20 \text{ MVA}, \frac{138 \text{ kV}}{20 \text{ kV}}, X = 10\%$$

$$T_{\Delta-Y} : 15 \text{ MVA}, \frac{138 \text{ kV}}{20 \text{ kV}}, X = 10\%$$

$$Z^{+,-}_{\text{Line1}} = j40 \ \Omega, \ Z^{+,-}_{\text{Line2}} = j20 \ \Omega, \ X^0_{\text{Line}} = 3X^+_{\text{Line}}$$

Difficulty level o Easy o Normal • Hard

3.8 If we have: $X^+_{33} = X^-_{33} = 0.0724$, $X^0_{33} = 0.0657$ and $Z_f = j0.1$. What is the current:

a. Single phase-to-ground.

b. Phase-to-phase, two-phase short circuit.

c. Two phase-to-ground short circuit.

d. Phase-to-phase, other single phase-to-ground.

Difficulty level o Easy o Normal • Hard

3.9 (See Figure 3.14). Which type of short circuit is it? Does the generator's and motor's star have a ground connection?

FIGURE 3.14 Question network 3.9.

Difficulty level o Easy •Normal o Hard

$$\hat{I}_1 \text{ or } \hat{I}_2 = \begin{cases} \hat{I}^+ = 2j-1 \\ \hat{I}^- = 3j \\ \hat{I}^0 = 4j \end{cases}, \quad \hat{I}_1 \text{ or } \hat{I}_2 = \begin{cases} \hat{I}^+ = 2j+1 \\ \hat{I}^- = j \\ \hat{I}^0 = 0 \end{cases}$$

3.8   DESCRIPTIVE ANSWERS TO THREE-PHASE UNSYMMETRICAL SHORT CIRCUIT

3.1   From equation 3.3:

$$\begin{bmatrix} \hat{I}_a^0 \\ \hat{I}_a^+ \\ \hat{I}_a^- \end{bmatrix} = \frac{1}{3}\begin{bmatrix} 1 & 1 & 1 \\ 1 & \alpha & \alpha^2 \\ 1 & \alpha^2 & \alpha \end{bmatrix}\begin{bmatrix} 1.6\angle 25 \\ 1.0\angle 180 \\ 0.9\angle 132 \end{bmatrix} = \frac{1}{3}\begin{bmatrix} 1.6\angle 25 - 1 + 0.9\angle 132 \\ 1.6\angle 25 - 1\angle 120 + 0.9\angle 12 \\ 1.6\angle 25 - 1\angle -120 + 0.9\angle 252 \end{bmatrix}$$

$$= \begin{bmatrix} 0.4512\angle 96.4529° \\ 0.9435\angle -0.0550° \\ 0.6024\angle 22.3157° \end{bmatrix}$$

$$\begin{cases} \hat{I}_a^0 = \hat{I}_b^0 = \hat{I}_c^0 = 0.4512\angle 96.4529° \\ \hat{I}_b^+ = \alpha^2\hat{I}_a^+ = 0.9435\angle -120.055° \quad ; \quad \hat{I}_c^+ = \alpha\hat{I}_a^+ = 0.9435\angle 119.945° \\ \hat{I}_b^- = \alpha\hat{I}_a^- = 0.6024\angle 142.3157° \quad ; \quad \hat{I}_c^- = \alpha^2\hat{I}_a^- = 0.6024\angle -97.6843° \end{cases}$$

3.2   We simply have:

$$\overset{I_a^{abc}}{\begin{bmatrix} \hat{I}_a \\ \hat{I}_b \\ \hat{I}_c \end{bmatrix}} = \overset{A}{\begin{bmatrix} 1 & 1 & 1 \\ 1 & \alpha^2 & \alpha \\ 1 & \alpha & \alpha^2 \end{bmatrix}}\overset{I_a^{0+-}}{\begin{bmatrix} \hat{I}_a^0 \\ \hat{I}_a^+ \\ \hat{I}_a^- \end{bmatrix}}, \quad \overset{I_a^{abc}}{\begin{bmatrix} \hat{I}_a \\ \hat{I}_b \\ \hat{I}_c \end{bmatrix}} = \overset{A^{New}}{\begin{bmatrix} 1 & 1 & 1 \\ \alpha^2 & \alpha & 1 \\ \alpha & \alpha^2 & 1 \end{bmatrix}}\overset{I_a^{+-0}}{\begin{bmatrix} \hat{I}_a^+ \\ \hat{I}_a^- \\ \hat{I}_a^0 \end{bmatrix}}$$

Then:

$$\Rightarrow \begin{bmatrix} \hat{I}_a^+ \\ \hat{I}_a^- \\ \hat{I}_a^0 \end{bmatrix} = \frac{1}{3} \begin{bmatrix} 1 & \alpha & \alpha^2 \\ 1 & \alpha^2 & \alpha \\ 1 & 1 & 1 \end{bmatrix} \begin{bmatrix} \hat{I}_a \\ \hat{I}_b \\ \hat{I}_c \end{bmatrix}, \quad \left(A^{New}\right)^{-1} = \frac{1}{3}\left(A^{New}\right)^*,$$

3.3 In the delta load we have:

$$\hat{I}_a + \hat{I}_b + \hat{I}_c = 0 \Rightarrow \hat{I}_c = -\left(\hat{I}_a + \hat{I}_b\right) = 18.028\angle153.69°$$

From equation 3.4:

$$I_a^{0+-} = A^{-1}I^{abc} \Rightarrow$$

$$\begin{bmatrix} \hat{I}_a^0 \\ \hat{I}_a^+ \\ \hat{I}_a^- \end{bmatrix} = \frac{1}{3}\begin{bmatrix} 1 & 1 & 1 \\ 1 & \alpha & \alpha^2 \\ 1 & \alpha^2 & \alpha \end{bmatrix} \begin{bmatrix} 10\angle30° \\ 15\angle-60° \\ 18.028\angle153.69° \end{bmatrix} = \begin{bmatrix} 0 \\ 13.962\angle41.932° \\ 4.662\angle-111.74° \end{bmatrix}$$

3.4 We have:

$$\hat{V}_A = \frac{4.8\text{ kV}}{\sqrt{3}}\angle0, \ \hat{V}_B = \frac{4.8\text{ kV}}{\sqrt{3}}\angle-120°, \ \hat{V}_C = \frac{4.8\text{ kV}}{\sqrt{3}}\angle120°$$

For 10 Ω resistance load:

$$\hat{V}_A = \frac{4.8\text{ kV}}{\sqrt{3}} \Rightarrow \hat{I}_a = \frac{4.8\text{ kV}}{10\sqrt{3}}\angle0 = 0.277\angle0 \text{ kA}$$

For j10 Ω reactance load:

$$\hat{I}_b = -\hat{I}_c = \frac{\hat{V}_B - \hat{V}_C}{jX_{bc}} = \frac{\dfrac{4.8\text{ kV}}{\sqrt{3}}(1\angle-120-1\angle120)}{10j} = 0.48\angle180 \text{ kA}$$

$$\begin{bmatrix} \hat{I}_a^0 \\ \hat{I}_a^+ \\ \hat{I}_a^- \end{bmatrix} = \frac{1}{3}\begin{bmatrix} 1 & 1 & 1 \\ 1 & \alpha & \alpha^2 \\ 1 & \alpha^2 & \alpha \end{bmatrix} \begin{bmatrix} 0.277\text{ kA} \\ -0.48\text{ kA} \\ 0.48\text{ kA} \end{bmatrix} = \begin{bmatrix} 92.33\angle0 \\ 292.11\angle-71.573° \\ 292.11\angle+71.573° \end{bmatrix} \text{A}$$

3.5

Method 1:

$$S = V^T I^* = \left[\begin{array}{ccc} \dfrac{4.8\,\text{kV} \angle 0}{\sqrt{3}} & \dfrac{4.8\,\text{kV} \angle -120}{\sqrt{3}} & \dfrac{4.8\,\text{kV} \angle 120}{\sqrt{3}} \end{array} \right] \left[\begin{array}{c} 277 \\ -480 \\ 480 \end{array} \right] =$$

$$= 2.429 \angle 71.573 \; \text{MVA} = 0.768 \; \text{MW} + j2.304 \; \text{MVAr}$$

Method 2:

$$P_R = \frac{V_R^2}{R} = \frac{\left(4.8\,\text{kV} / \sqrt{3}\right)^2}{10\,\Omega} = 0.768 \; \text{MW}$$

$$Q_X = \frac{V_X^2}{X} = \frac{(4.8\,\text{kV})^2}{10\,\Omega} = 2.304 \; \text{MVAr}$$

Method 3: (From equation 3.10)

$$S_{3ph} = 3\left(V_a^{0+-}\right)^T \left(I_a^{0+-}\right)^* = 3\left(\hat{V}_a^0\,\hat{I}_a^{0*} + \hat{V}_a^+\,\hat{I}_a^{+*} + \hat{V}_a^-\,\hat{I}_a^{-*}\right)$$

Symmetrical three-phase source:

$$\hat{V}_a^+ = \frac{4.8\,\text{kV}}{\sqrt{3}}, \; \hat{V}_a^- = \hat{V}_a^0 = 0$$

$$\Rightarrow S_{3ph} = 3\left(\hat{V}_a^+\,\hat{I}_a^{+*}\right) = 3\left(\frac{4.8\,\text{kV}}{\sqrt{3}}\right)(292.11 \angle 71.573°) = 2.429 \angle 71.573° \; \text{MVA}$$

3.6 The transformer connections do not impact the positive and negative sequence networks. The negative sequence network mirrors the positive sequence network but lacks voltage sources. See Figure 3.15. Neither the positive nor the negative network exhibits the neutral impedance.

In the zero network, neutral impedance (Z_n) appears three times ($3Z_n$). See Figure 3.16.

3.7 We define $S_b^{\text{new}} = 20$ MW and base voltage in generator 1 is $V_b^{\text{new}} = 20\,\text{kV}$. Then the impedances in the new base are computed using the following formula.

$$Z^{\text{new}} = Z^{\text{old}} \left(\frac{S_b^{\text{new}}}{S_b^{\text{old}}}\right) \left(\frac{V_b^{\text{old}}}{V_b^{\text{new}}}\right)^2$$

$$\Rightarrow \left(X_g^+\right)^{\text{new}} = \left(X_g^-\right)^{\text{new}} = 0.2 \left(\frac{13}{20}\right)^2 = 0.0845 \; \text{pu}$$

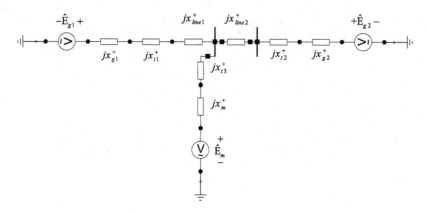

FIGURE 3.15 Answer 1 network 3.6.

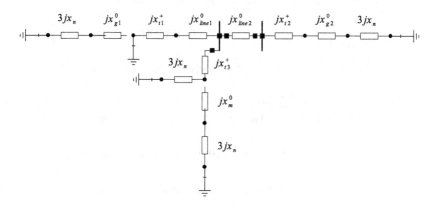

FIGURE 3.16 Answer 2 network 3.6.

$$\Rightarrow \left(X_g^n\right)^{\text{new}} = 0.05\left(\frac{13}{20}\right)^2 = 0.0211 \text{ pu}$$

$$\Rightarrow \left(X_g^0\right)^{\text{new}} = 0.08\left(\frac{13}{20}\right)^2 = 0.0338 \text{ pu}$$

$$T_{yy} = T_2 : \left(X_{T_2}^{+,-,0}\right)^{\text{new}} = 0.1\left(\frac{20}{20}\right)\left(\frac{20}{20}\right)^2 = 0.1 \text{ pu}$$

$$T_{\Delta y} = T_1 = T_3 : \left(X_{T_1,T_3}^{+,-,0}\right)^{\text{new}} = 0.1\left(\frac{20}{15}\right)\left(\frac{20}{20}\right)^2 = 0.133 \text{ pu}$$

$$M : \left(X_m^{+,-}\right)^{\text{new}} = 0.2\left(\frac{20}{30}\right)\left(\frac{13.8}{20}\right)^2 = 0.0635 \text{ pu}$$

$$M: \left(X_m^0\right)^{new} = 0.08\left(\frac{20}{30}\right)\left(\frac{13.8}{20}\right)^2 = 0.0254 \text{ pu}$$

$$M: \left(X_m^n\right)^{new} = 0.05\left(\frac{20}{30}\right)\left(\frac{13.8}{20}\right)^2 = 0.0159 \text{ pu}$$

$$\text{Line}: Z_{base}^{new} = \frac{V_b^2}{S_b} = \frac{(138\,\text{kV})^2}{20\,\text{MVA}} = 952.2\,\Omega \Rightarrow \left(X_{Line1}^{+,-}\right)^{new} = \frac{40}{952.2} = 0.042 \text{ pu}$$

$$\Rightarrow \left(X_{Line2}^{+,-}\right)^{new} = \frac{20}{952.2} = 0.021 \text{ pu}$$

Figure 3.17 is a positive sequence.
The formula for bus 3's Thévenin impedance is as follows.

$$X_{33}^+ = (0.042 + 0.133 + 0.0845) \,\|\, (0.133 + 0.0635) \,\|\, (0.021 + 0.1 + 0.0845)$$

$$X_{33}^+ = X_{33}^- = 0.2595 \,\|\, 0.1965 \,\|\, 0.2055 = 0.0724 \text{ pu}$$

Figure 3.18 is a zero sequence.
The formula for bus 3's Thévenin impedance is as follows.

$$X_{33}^0 = (3(0.042) + 0.133) \,\|\, (0.133) \,\|\, (3(0.021) + 0.1 + 0.0338 + 3(0.0211)) \Rightarrow$$

$$X_{33}^0 = (0.2590) \,\|\, (0.133) \,\|\, (0.2601) = 0.0657 \text{ pu}$$

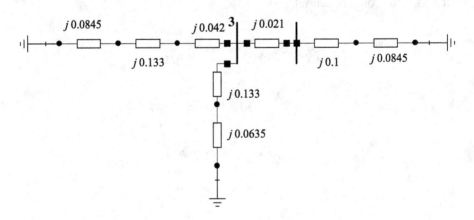

FIGURE 3.17   Answer 1 network 3.7.

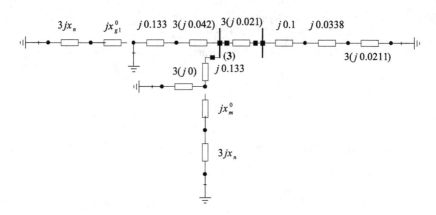

FIGURE 3.18 Answer 2 network 3.7.

3.8 For each short circuit, we use its own equivalent circuit.

3.8.a Single phase-to-ground. From equation 3.20:

$$\frac{\hat{I}_f}{3} = \frac{\hat{E}_a}{Z^+ + Z^- + Z^0 + 3Z_f} = \frac{1}{2(j0.0724) + j0.0657 + 3(j0.1)}$$

$$\Rightarrow \hat{I}_f = -j5.877$$

3.8.b Phase-to-phase, two-phase short circuit. See Figure 3.19. From equation 3.23:

$$\hat{I}_a^+ = \frac{j\sqrt{3}}{3}\hat{I}_f$$

$$\hat{I}_a^+ = \frac{1}{2(j0.0724) + j0.1} = -j4.085$$

$$\hat{I}_a^+ = -\hat{I}_a^- = \frac{j\sqrt{3}}{3}\hat{I}_f \Rightarrow$$

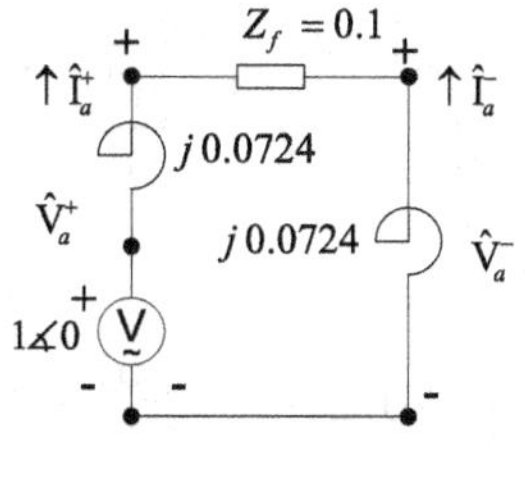

FIGURE 3.19 Answer 1 network 3.8.

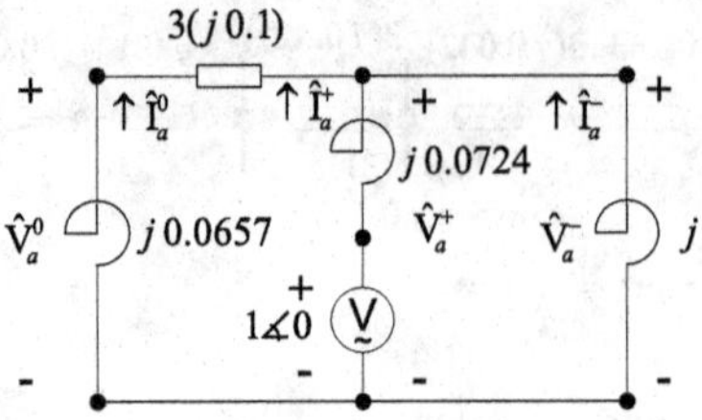

FIGURE 3.20 Answer 2 network 3.8.

$$\hat{I}_a^+ = -\hat{I}_a^- = \frac{j\sqrt{3}}{3}\hat{I}_f \Rightarrow \hat{I}_f = \frac{3}{j\sqrt{3}}(-j4.085) = -7.0754 \text{ pu}$$

3.8.c Two phase-to-ground short circuit. See Figure 3.20. From equation 3.31:

$$\hat{I}_a^0 = \frac{\hat{I}_f}{3}$$

$$\hat{I}_a^+ = \frac{1}{j(0.0724 + (0.0724 \parallel (0.3 + 0.0657)))} = \frac{1}{j(0.0724 + 0.0604)} \Rightarrow$$

$$\hat{I}_a^+ = -j7.528 \text{ pu}$$

$$\hat{I}_a^0 = -j7.528\left(\frac{-0.0724}{0.0724 + 0.3657}\right) = j1.2441$$

$$\Rightarrow \hat{I}_f = 3\hat{I}_a^0 = j3.7323 \text{ pu}$$

3.8.d Phase-to-phase, other single phase-to-ground. From equations 3.38 and 3.41: $\frac{2}{3}\hat{I}_{fi} = 2\hat{I}_a^0$. It is assumed that $Z_{f1} = Z_{f2} = 0$

$$\hat{I}_a^+ = \frac{\hat{E}}{jx^+ + \left(jx^- \parallel \dfrac{jx^0}{4}\right)}, \quad \hat{I}_a^- = \frac{-\dfrac{jx^0}{4}}{jx^- + \dfrac{jx^0}{4}}\hat{I}_a^+, \quad 2\hat{I}_a^0 = \frac{jx^-}{jx^- + \dfrac{jx^0}{4}}\hat{I}_a^+$$

$$\hat{I}_a^+ = \frac{1}{j0.0724 + \left(j0.0724 \parallel \dfrac{j0.0657}{4}\right)} = -j11.6567 \text{ pu}$$

$$\hat{I}_a^- = \frac{-\dfrac{j0.0657}{4}}{j0.0724 + \dfrac{j0.0657}{4}}(-j11.6567) = j2.1555 \text{ pu}$$

$$2\hat{I}_a^0 = \frac{j0.0724}{j0.0724 + \dfrac{j0.0657}{4}}(-j11.6567) = -j9.5012 \text{ pu}$$

$$\hat{I}_{f1} = \frac{3}{2}(2\hat{I}_a^0) = -j14.252 \text{ pu}$$

3.9 $\hat{I}_{SC} = \hat{I}_1 + \hat{I}_2 \Rightarrow$

$$\begin{cases} \hat{I}_{SC}^+ = 4j \\[2mm] \hat{I}_{SC}^- = j + 3j = 4j \Rightarrow \\[2mm] \hat{I}_{SC}^0 = 4j + 0 = 4j \end{cases}$$

$$\Rightarrow \hat{I}_{SC}^+ = \hat{I}_{SC}^- = \hat{I}_{SC}^0 \Rightarrow$$

A single-phase short circuit to the ground has occurred. Since the transformer's connection prevents the zero component from passing, the I_1 current is the only current that lacks it. I_2 is the current that contains a zero component. Since the motor has allowed current to flow, the ground is connected to the center of the motor's star. It is impossible to talk about the generator's star center, but based on past experience, it needs to be connected to the ground.

NOTE

1 Line transposition refers to the process of altering the physical arrangement of conductors in a transmission line to mitigate certain undesirable effects such as unequal electrical characteristics between phases or to balance mutual coupling effects. Physically it is done through repositioning the conductors at regular intervals along the transmission line. For example, switching the positions of conductors in a bundle or altering the sequence in which conductors are suspended on transmission towers.

4 Transient Stability Analysis

Part One: Lesson Summary

4.1 INTRODUCTION

The equal area criterion, stability limit curve, critical clearing angle, maximum power, and numerical solution of the dynamic equation are all reviewed in this chapter along with the transient stability analysis equations. The most crucial dynamic limit for power system planning and operation is transient stability. Determining transient stability limits and enhancing transient stability techniques are the two main subgroups of transient stability issues. Transient stability limit determination is the main topic of this chapter. The equal area criterion method is the easiest way to figure out the system's critical clearing time. The simulation method, energy method, Lyapunov, neural networks, etc., are being used in real problems and large networks. This chapter's problems are made in a way that makes using a computer unnecessary.

4.2 BASIC CONCEPTS

4.2.1 GENERAL TYPES OF STABILITY

1. Static stability

$$\dot{X} = \frac{dX}{dt} = 0$$

2. Dynamic stability

$$\text{2-1) Small signal } \dot{X} = AX$$

$$\text{2-2) Large signal } \dot{X} = f(X)$$

2-1. Small signal: small variations and linearization around the operating point.
Static instability or **voltage instability** results when the load flow is disrupted. Voltage collapse is the same as the voltage instability limit.

Solution method:	Gauss-Seidel (GS), Newton Raphson (NR), or Newton Raphson Seidel (NRS).

DOI: 10.1201/9781003506751-4

2-2. Large signal: large changes, for example, short circuit, removal of loads, outage of a line, outage of a transformer

---

Solution method: Solving the nonlinear differential equation with Euler's method, modified Euler's method, the Runge-Kutta method

---

4.2.2   DEFINITION OF THE INFINITE BUS

At an **infinite bus**, both the bus frequency and voltage magnitude are fixed. Large (infinite) loads and (infinite) generators are present in the network that lies behind this bus.

4.2.3   REMINDER

The cylindrical rotor generator output power:

$$P_e = \frac{V_0 V_\infty}{X} \operatorname{Sin}\delta, \ \operatorname{Max}(P_e) = \frac{V_0 V_\infty}{X} \tag{4.1}$$

4.2.4   TRANSIENT STABILITY DEFINITION

The simplest definition of transient stability is: "Keeping synchronism in the face of a fault". If it is capable of returning to a synchronous state, it is stable; if not, it is unstable.

4.2.5   IMPORTANT QUESTIONS IN TRANSIENT STABILITY

- For how long can a short circuit persist before the system becomes unstable?
- The system remains stable for a given short circuit time?

4.2.6   TRANSIENT STABILITY IMPROVEMENT METHODS

- The use of resistance is called **braking resistors.**
- Using **SMES** (Superconductive Magnetic Energy Storage), magnetically storing electric energy in a superconductor coil.
- **Quick disconnection** of faults by quickly disconnecting circuit breakers or switches.

4.3   TRANSIENT STABILITY ANALYSIS EQUATIONS

Newton's second law, linear motion:

$$\Sigma F = M\,a$$

Newton's second law, uniform circular motion:

$$\Sigma T = T_m - T_e = T_a = J\frac{d^2\theta_m}{dt^2}$$

Inertia constant in (MWs):

$$M \cong J\omega_n = \frac{2W_k}{\omega_n}$$

Inertia constant in pu (s) or (MWs/MVA):

$$H \triangleq \frac{W_k}{S_b}$$

Swing or dynamic equations of a synchronous machine in pu:

$$\frac{H}{\pi f^0}\frac{d^2\delta}{dt^2} = P_m - P_e \qquad (4.2)$$

where:

P_m: The turbine's mechanical power
P_e: Generator electrical output power
δ: Electric angle (radians)
f^0: Operating point frequency (synchronous frequency)

4.4   SYNCHRONOUS GENERATOR MODEL

4.4.1   CYLINDRICAL ROTOR (NORMAL & TRANSIENT)

Note that a cylindrical rotor is equivalent to a salient pole rotor by ignoring the saliency. One of the following impedances is regarded as the machine's internal impedance, depending on the short-circuit time.

$$X = X_S, \text{ or } X_d \quad (\text{in normal state}) \qquad (4.3)$$

$$X = X_d', \text{ or } X_d'' \quad (\text{in the transient state}) \qquad (4.4)$$

Then:

$$\hat{V} = V\angle 0, \ \hat{E} = E\angle\delta \Rightarrow P_e = \frac{E\,V}{X}\operatorname{Sin}\delta \qquad (4.5)$$

Example 4.1

Find the power equations of a cylindrical rotor synchronous machine connected by a line to an infinite bus (Figure 4.1).

Power equations:

$$P_e = \frac{EV}{X}\operatorname{Sin}(\delta - \delta_1) = \frac{VV_\infty}{X_L}\operatorname{Sin}(\delta_1 - 0) = \frac{EV_\infty}{X + X_L}\operatorname{Sin}(\delta - 0) \qquad (4.6)$$

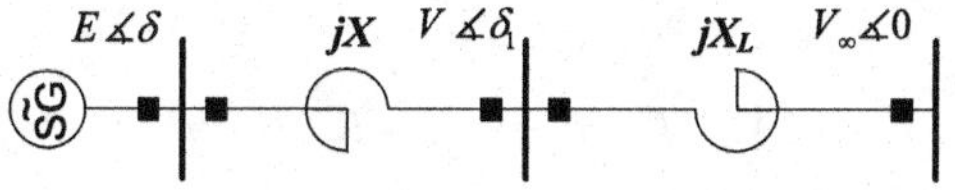

$$E\angle\delta \qquad jX \quad V\angle\delta_1 \qquad jX_L \quad V_\infty\angle0$$

FIGURE 4.1   A machine connected (through a line) to an infinite bus.

$$\Rightarrow \text{Max } (P_e) = \frac{EV_\infty}{X + X_L} \qquad (4.7)$$

Due to the requirement for constant voltages, maximum power is only true for the final relation. For (X) see equations 4.3 or 4.4.

4.4.2   SALIENT POLE ROTOR (NORMAL)

The following is an expression for the equations derived from the synchronous generator's phasor diagram.

$$\hat{E} = \hat{V} + r\hat{I}_a + jX_d\hat{I}_d + jX_q\hat{I}_q \qquad (4.8)$$

or:

$$\hat{E} = \hat{V} + r\hat{I}_a + jX_q\hat{I}_a + j(X_d - X_q)\hat{I}_d \qquad (4.9)$$

Example 4.2

(See Figure 4.2). Find (δ) if (V, I, θ, X_d, X_q) are known.
   Method 1 $(r = 0)$:

$$\begin{cases} I_q = I_a \cos(\delta + \theta) & (1) \\ I_d = I_a \sin(\delta + \theta) & (2) \end{cases} \qquad (4.10)$$

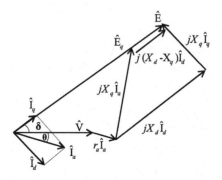

FIGURE 4.2   A synchronous generator's phasor diagram.

$$\begin{cases} V \sin \delta = X_q I_q & (3) \\ V \cos \delta + X_d I_d = E & (4) \end{cases} \qquad (4.11)$$

$$\xrightarrow{1,3} V \sin\delta = X_q I_a \cos(\delta+\theta) = X_q I_a \left(\cos\delta \cos\theta - \sin\delta \sin\theta\right) \Rightarrow$$

$$\xrightarrow{\div \cos\delta} \tan\delta\left(V + X_q I_a \sin\theta\right) = X_q I_a \cos\theta \Rightarrow$$

$$\Rightarrow \delta = \tan^{-1}\left(\frac{X_q I_a \cos\theta}{V + X_q I_a \sin\theta}\right) \qquad (4.12)$$

Method 2:

$$\hat{E} = \hat{V} + r\hat{I}_a + jX_d\hat{I}_d + jX_q\left(\hat{I}_a - \hat{I}_d\right) \Rightarrow$$

$$\Rightarrow \hat{E} = \hat{V} + r\hat{I}_a + jX_q\hat{I}_a + j\left(X_d - X_q\right)\hat{I}_d$$

$$\hat{E}_q \triangleq \hat{V} + r\hat{I}_a + jX_q\hat{I}_a \Rightarrow \hat{E} = \hat{E}_q + j(X_d - X_q)\hat{I}_d \qquad (4.13)$$

$$\hat{E}_q = E_q \angle \delta \Rightarrow \delta = \angle\left(\hat{E}_q\right) \qquad (4.14)$$

4.4.3 Salient Pole Rotor (Transient)

(See Figure 4.2) If (X_d) becomes (X_d') and $(\hat{E})$ becomes $(\hat{E}')$, the transient state equation is obtained.

$$\hat{E}' = \hat{V} + r\hat{I}_a + jX_d'\hat{I}_d + jX_q\hat{I}_q \qquad (4.15)$$

or:

$$\hat{E}' = \hat{V} + r\hat{I}_a + jX_q\hat{I}_a + j(X_d' - X_q)\hat{I}_d \qquad (4.16)$$

and:

$$E - E' = \left(X_d - X_d'\right)I_d \qquad (4.17)$$

Power equations:

$$\text{Normal}: P_e = \frac{EV}{X_d}\sin\delta + \frac{V^2}{2}\left(\frac{1}{X_q} - \frac{1}{X_d}\right)\sin 2\delta \qquad (4.18)$$

$$\text{Transient}: P_e = \frac{E'V}{X_d'}\sin\delta + \frac{V^2}{2}\left(\frac{1}{X_q} - \frac{1}{X_d'}\right)\sin 2\delta \qquad (4.19)$$

4.4.4   THE LINE REACTANCE EFFECT

X_L is the line reactance between the generator and the bus infinite. Then we define:

$$\tilde{X}_d \triangleq X_d + X_L,\ \tilde{X}_q \triangleq X_q + X_L,\ \tilde{X}_d' \triangleq X_d' + X_L,\ \tilde{X}_S \triangleq X_S + X_L \qquad (4.20)$$

Therefore, $(\tilde{X})$ is used in place of (X) in all power and voltage equations.

4.5   THE EQUAL AREA CRITERION FOR TRANSIENT STABILITY

When a generator is connected to an infinite bus, this technique is employed.
   Proof of the equal area criterion: From equation 4.2

$$\Rightarrow \frac{H}{\pi f^0}\frac{d^2\delta}{dt^2} = P_m - P_e \quad \Rightarrow \quad \frac{d^2\delta}{dt^2} = \frac{\pi f^0}{H}(P_m - P_e)$$

$$\xrightarrow{2\frac{d\delta}{dt}}\ 2\frac{d\delta}{dt}\frac{d^2\delta}{dt^2} = \frac{2\pi f^0}{H}(P_m - P_e)\frac{d\delta}{dt} \Rightarrow$$

$$\Rightarrow \frac{d}{dt}\left(\left(\frac{d\delta}{dt}\right)^2\right) = \frac{2\pi f^0}{H}(P_m - P_e)\frac{d\delta}{dt}$$

$$\Rightarrow d\left(\left(\frac{d\delta}{dt}\right)^2\right) = \frac{2\pi f^0}{H}(P_m - P_e)\,d\delta \Rightarrow$$

$$\Rightarrow \left(\frac{d\delta}{dt}\right)^2 = \frac{2\pi f^0}{H}\int_{\delta^0}^{\delta}(P_m - P_e)\,d\delta$$

Stability condition:

$$\exists \delta : \frac{d\delta}{dt} = 0 \Rightarrow \int_{\delta^0}^{\delta}(P_m - P_e)\,d\delta = 0 \qquad (4.21)$$

The integral of the difference between mechanical and electrical power must be zero since they change during the fault. As a result, the positive and negative areas should be equal, as indicated by the area under the curve being zero.

4.6   TRANSIENT STABILITY ANALYSIS WITH SHORT CIRCUIT

The mechanical power (P_m) is assumed to be constant in this case.
   Electric power is required in three modes.

1. Pre-fault system: $P_{e1}(\delta)$
2. Faulty system: $P_{e2}(\delta)$
3. Post-fault system: $P_{e3}(\delta)$

Stability requires that the energy taken in mode 2 be transferred to the grid within a certain time frame.

4.6.1 Stable System, Maximum Angle

According to Figure 4.3, the maximum delta (δ_{max}) is unknown, and the rest of the parameters are known. The equal area criterion for transient stability is written as follows.

$$\text{Acceleration area}: A_1 = A_{acc} = \int_{\delta_0}^{\delta_{cl}} \left(P_m - P_{e_2}(\delta)\right) d\delta \tag{4.22}$$

$$\text{Deceleration area}: A_2 = A_{dec} = \int_{\delta_{cl}}^{\delta_{max}} \left(P_{e_3}(\delta) - P_m\right) d\delta \tag{4.23}$$

$$\text{Stability condition}: \left(A_1 = A_2\right) \text{ or } \left(A_{acc} = A_{dec}\right) \tag{4.24}$$

The equation above can be used to determine the maximum delta (δ_{max}).

4.6.2 Stable System, Minimum Angle

According to Figure 4.4, the minimum delta (δ_{min}) is unknown, and the rest of the parameters are known. The equal area criterion for transient stability is written as follows.

$$\text{Acceleration area}: A_3 = A_{acc} = \int_{\delta_1}^{\delta_{max}} \left(P_{e_3}(\delta) - P_m\right) d\delta \tag{4.25}$$

$$\text{Deceleration area}: A_4 = A_{dec} = \int_{\delta_{min}}^{\delta_1} \left(P_m - P_{e_3}(\delta)\right) d\delta \tag{4.26}$$

$$\text{Stability condition}: \left(A_3 = A_4\right) \text{ or } \left(A_{acc} = A_{dec}\right) \tag{4.27}$$

The equation above can be used to determine the minimum delta (δ_{min}).

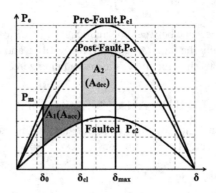

FIGURE 4.3   Stable system, maximum angle.

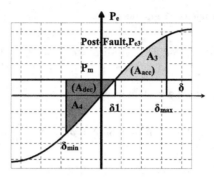

FIGURE 4.4 Stable system, minimum angle.

4.6.3 STABILITY LIMIT CURVE, CRITICAL CLEARING ANGLE

(See Figure 4.5) In this case, the short circuit continues until the system reaches the stability limit. In other words, if the short circuit time is added as much as an Epsilon, the network becomes unstable. According to Figure 4.5, the critical clearing angle (δ_{cr}) is unknown, and the rest of the parameters are known. The equal area criterion for transient stability is written as follows.

$$A_{\text{acc}} = A_{\text{dec}} \Rightarrow \int_{\delta_0}^{\delta_{cr}} \left(P_m - P_{e_2}(\delta) \right) d\delta = \int_{\delta_{cr}}^{\delta_{\text{max}}} \left(P_{e_3}(\delta) - P_m \right) d\delta \qquad (4.28)$$

Example 4.3

(See Figure 4.5) Find the critical clearing angle (δ_{cr}) if the generator is a cylindrical rotor.

We define:

1. Pre-fault system: $P_{e_1}(\delta) = P_1 \, \mathrm{Sin}\,\delta$
2. Faulty system: $P_{e2}(\delta) = P_2 \, \mathrm{Sin}\,\delta$
3. Post-fault system: $P_{e_3}(\delta) = P_3 \, \mathrm{Sin}\,\delta$

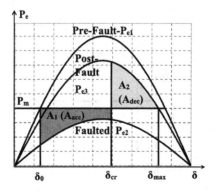

FIGURE 4.5 Stable system, critical clearing angle.

Stable equilibrium angle of Pre-fault system:

$$\delta_0 : P_m = P_{e_1}(\delta) = P_1 \, \text{Sin}\,\delta \Rightarrow \delta_0 = \text{Sin}^{-1}\left(\frac{P_m}{P_1}\right) \qquad (4.29)$$

Stable equilibrium angle of a Post-fault system:

$$\delta_1 : P_m = P_{e_3}(\delta) = P_3 \, \text{Sin}\,\delta \Rightarrow \delta_1 = \text{Sin}^{-1}\left(\frac{P_m}{P_3}\right) \qquad (4.30)$$

Unstable equilibrium angle of a Post-fault system:

$$\delta_{\max} : P_m = P_3 \, \text{Sin}\,\delta \Rightarrow \delta_{\max} = \pi - \delta_1 \qquad (4.31)$$

$$A_{\text{acc}} = A_{\text{dec}} \Rightarrow \int_{\delta_0}^{\delta_{cr}}\left(P_m - P_2 \, \text{Sin}\,\delta\right)d\delta = \int_{\delta_{cr}}^{\delta_{\max}}\left(P_3 \, \text{Sin}\,\delta - P_m\right)d\delta \Rightarrow$$

$$P_m\left(\delta_{cr} - \delta_0\right) + P_2\left(\text{Cos}\,\delta_{cr} - \text{Cos}\,\delta_0\right) = -P_3\left(\text{Cos}\,\delta_{\max} - \text{Cos}\,\delta_{cr}\right) - P_m\left(\delta_{\max} - \delta_{cr}\right)$$

$$\Rightarrow \left(P_2 - P_3\right)\text{Cos}\,\delta_{cr} = P_2\,\text{Cos}\,\delta_0 - P_3\,\text{Cos}\,\delta_{\max} - P_m\left(\delta_{\max} - \delta_0\right)$$

$$\Rightarrow \delta_{cr} = \text{Cos}^{-1}\left(\frac{P_2\,\text{Cos}\,\delta_0 - P_3\,\text{Cos}\,\delta_{\max} - P_m\left(\delta_{\max} - \delta_0\right)}{\left(P_2 - P_3\right)}\right) \qquad (4.32)$$

4.7 TRANSIENT STABILITY ANALYSIS WITH CHANGES IN MECHANICAL POWER

4.7.1 STABLE STATE

(See Figure 4.6) The method of equal area criterion can be used to calculate the maximum angle by changing the mechanical power.

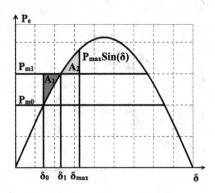

FIGURE 4.6 Stable system, transient stability analysis with changes in mechanical power.

Stable equilibrium angle of pre-change system:

$$\delta_0 : P_{m0} = P_{max} \, \text{Sin} \, \delta \Rightarrow \delta_0 = \text{Sin}^{-1}\left(\frac{P_{m0}}{P_{max}}\right) \tag{4.33}$$

Stable equilibrium angle of post-change system:

$$\delta_1 : P_{m1} = P_{max} \, \text{Sin} \, \delta \Rightarrow \delta_1 = \text{Sin}^{-1}\left(\frac{P_{m1}}{P_{max}}\right) \tag{4.34}$$

The equal area criterion for transient stability is written as follows.

$$A_{acc} = A_{dec} \Rightarrow \int_{\delta_0}^{\delta_1}\left(P_{m1} - P_{max} \, \text{Sin} \, \delta\right) d\delta = \int_{\delta_1}^{\delta_{max}}\left(P_{max} \, \text{Sin} \, \delta - P_{m1}\right) d\delta$$

$$\Rightarrow P_{m1}\left(\delta_1 - \delta_0\right) + P_{max}\left(\text{Cos} \, \delta_1 - \text{Cos} \, \delta_0\right) = -P_{max}\left(\text{Cos} \, \delta_{max} - \text{Cos} \, \delta_1\right) - P_{m1}\left(\delta_{max} - \delta_1\right)$$

$$\Rightarrow P_{max} \, \text{Cos} \, \delta_{max} + P_{m1}\delta_{max} + \left(-P_{m1}\delta_0 - P_{max} \, \text{Cos} \, \delta_0\right) = 0 \tag{4.35}$$

Iterative methods, such as Newton's, can solve equation 4.35. If (r) is the step and ($r = 0$) is the initial estimate, then we have:

$$f(x) = 0 \Rightarrow x^{(r+1)} = x^{(r)} - \frac{f(x^{(r)})}{f'(x^{(r)})} \tag{4.36}$$

4.7.2 MAXIMUM POWER

(See Figure 4.7) The method of equal area criteria can be used to calculate the maximum change in mechanical power.

See Figure 4.7.

$$\delta_0 = \text{Sin}^{-1}\left(\frac{P_{m0}}{P_{max}}\right), \quad \delta_1 = \text{Sin}^{-1}\left(\frac{P_{m1}}{P_{max}}\right), \quad \delta_{max} = \pi - \delta_1$$

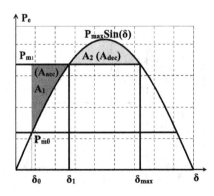

FIGURE 4.7 Maximum power, transient stability analysis with change in mechanical power.

$$A_{acc} = A_{dec} \Rightarrow \int_{\delta_0}^{\delta_1} \left(P_{m1} - P_{max} \ Sin\,\delta \right) d\delta = \int_{\delta_1}^{\delta_{max}} \left(P_{max} \ Sin\,\delta - P_{m1} \right) d\delta$$

$$\Rightarrow P_{m1}\left(\delta_1 - \delta_0\right) + P_{max}\left(Cos\,\delta_1 - Cos\,\delta_0\right) = -P_{max}\left(Cos\,\delta_{max} - Cos\,\delta_1\right) - P_{m1}\left(\delta_{max} - \delta_1\right)$$

$$\Rightarrow -P_{m1}\delta_0 - P_{max}\ Cos\,\delta_0 + P_{max}\ Cos\,\delta_{max} + P_{m1}\delta_{max} = 0$$

We have:

$$P_{m1} = P_{max}\ Sin\,\delta_1$$

$$\Rightarrow -(P_{max}\ Sin\,\delta_1)\delta_0 - P_{max}\ Cos\,\delta_0 + P_{max}\ Cos(\pi - \delta_1) + (P_{max}\ Sin\,\delta_1)(\pi - \delta_1) = 0$$

$$\Rightarrow (Sin\,\delta_1)(\pi - \delta_0 - \delta_1) - Cos\,\delta_0 - Cos(\delta_1) = 0 \qquad (4.37)$$

Newton's method can solve equation 4.37, which is a nonlinear equation in terms of (δ_1).

4.8 SPECIAL MODES

During a short circuit, the transmitted power is zero and the network post and pre-fault are the same. For instance, the bus will self-correct if a short circuit occurs. Determining the maximum angle in a stable state is depicted in Figure 4.8. Determining the critical clearing angle is depicted in Figure 4.9.

4.9 THE NUMERICAL SOLUTION OF THE DYNAMIC EQUATION

4.9.1 ZERO ELECTRIC POWER

See Section 4.8 (special modes). From equation 4.2:

$$P_m - P_e\left(\delta\right) = \frac{H}{\pi f^0}\frac{d^2\delta}{dt^2}, \quad P_e = 0 \Rightarrow P_m = \frac{H}{\pi f^0}\frac{d^2\delta}{dt^2} \Rightarrow \frac{d^2\delta}{dt^2} = \frac{\pi f^0}{H}P_m$$

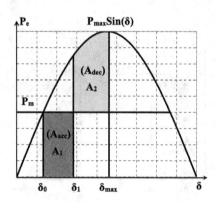

FIGURE 4.8 Special modes, the maximum angle in a stable state.

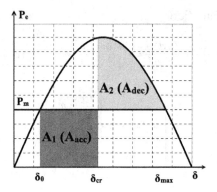

FIGURE 4.9 Special modes, the critical clearing angle.

$$\Rightarrow \frac{d\delta}{dt} = \frac{\pi f^0}{H} P_m t + \omega_0 \quad \overset{\omega_0=0}{\Rightarrow} \quad \delta = \frac{\pi f^0}{2H} P_m t^2 + \delta_0$$

$$\Rightarrow \delta_{cl} = \frac{\pi f^0}{2H} P_m t_{cl}^2 + \delta_0 \Rightarrow t_{cl} = \sqrt{\frac{2H}{\pi f^0 P_m}(\delta_{cl} - \delta_0)} \qquad (4.38)$$

If the clearing angle is known, the clearing time can be computed using equation 4.38.

4.9.2 NON-ZERO ELECTRIC POWER

4.9.2.1 Euler's Method

$$\frac{dx}{dt} = \dot{x} = f(x); \quad \frac{dx}{dt} \cong \frac{\Delta x}{\Delta t} = f(x); \quad x_1 = x_0 + \Delta x_0$$

$$x_1 = x_0 + f(x_0)\Delta t \quad \Rightarrow \quad x_{i+1} = x_i + f(x_i)\Delta t$$

4.9.2.2 Modified Euler's Method

$$\text{Estimate}: x_1^c = x_0 + f(x_0)\Delta t$$

$$\text{Correction}: x_1^P = x_0 + \left(\frac{f(x_0) + f(x_1^c)}{2}\right)\Delta t$$

$$\Rightarrow \begin{cases} x_{i+1}^c = x_i + f(x_i)\Delta t \\ \\ x_{i+1}^P = x_i + \left(\frac{f(x_i) + f(x_{i+1}^c)}{2}\right)\Delta t \end{cases} \qquad (4.39)$$

4.9.2.3 Dynamic Equation

$$P_m - P_e(\delta) = \frac{H}{\pi f^0}\frac{d^2\delta}{dt^2} \Rightarrow \frac{d^2\delta}{dt^2} = \frac{\pi f^0}{H}\left(P_m - P_e(\delta)\right) \triangleq P_a(\delta)$$

$$\Rightarrow \dot{x} = f(x) \Rightarrow \begin{cases} \dfrac{d\delta}{dt} = \omega \\[2mm] \dfrac{d\omega}{dt} = P_a(\delta) \end{cases}$$

We define:

$$x = \begin{bmatrix} \delta \\ \omega \end{bmatrix}, \quad f(x) = \begin{bmatrix} \omega \\ P_a(\delta) \end{bmatrix}$$

$$\begin{cases} \begin{bmatrix} \delta_{i+1}^c \\ \omega_{i+1}^c \end{bmatrix} = \begin{bmatrix} \delta_i \\ \omega_i \end{bmatrix} + \begin{bmatrix} \omega_i \\ P_a(\delta_i) \end{bmatrix}\Delta t \\[6mm] \begin{bmatrix} \delta_{i+1}^P \\ \omega_{i+1}^P \end{bmatrix} = \begin{bmatrix} \delta_i \\ \omega_i \end{bmatrix} + \frac{1}{2}\begin{bmatrix} \omega_i + \omega_{i+1}^c \\ P_a(\delta_i) + P_a(\delta_{i+1}^c) \end{bmatrix}\Delta t \end{cases} \tag{4.40}$$

Note: The answer can be more accurate with the smaller Δt, but the computations will take longer.

Part Two: Answer-Question

4.10   TWO-CHOICE QUESTIONS (YES/NO)

1. A generator will become unstable faster if its inertia constant is smaller.
2. The variables in the transient stability analysis exhibit a minimal variation.
3. Using SMES is a method to enhance transient stability.
4. When the bus is short-circuited, there is no transmitted power.
5. Interconnecting two networks generally enhances their overall transient stability.
6. The system eventually returns to its initial stable equilibrium point, if it is stable.
7. The relationship ($P = V_1V_2\mathrm{Sin}(\delta)/X$) holds true if there is no resistance.
8. The relationship ($\mathrm{Max}(P) = V_1V_2/X$) holds true if there is no resistance.
9. The transmitted power is increased by the line's impedance if the generator is linked to the infinite bus via a line.
10. The relation ($\pi - \delta$) yields the unstable equilibrium point in the salient pole generator.

11. This relationship $(\text{Max}(P_{e1}) \geq \text{Max}(P_{e2}))$ always holds true.
12. This relationship $(\text{Max}(P_{e2}) > \text{Max}(P_{e3}))$ always holds true.
13. This relationship $(\text{Max}(P_{e3}) > 0)$ always holds true.
14. The system is unquestionably stable if the acceleration area is larger than the deceleration area.
15. Accelerator and decelerator areas are always positive.
16. The system is definitely unstable if the generator angle reaches the unstable equilibrium point.
17. A system may be unstable in theory but stable in reality.
18. A system may be stable in theory but unstable in reality.
19. Transient stability can be achieved using linear control theory.
20. A system definitely unstable if a short circuit remains on the bus for a long period of time.
21. It is also possible to use the method of equal areas for large networks.
22. After a fault, the power delivered to the network is less than the maximum transmission power.
23. The network is stable if the clearing angle exceeds the critical angle.
24. Static stability analysis is simpler compared to dynamic stability analysis.
25. Static stability is less important than dynamic stability.
26. The term transient stability refers to the voltage's ability to remain stable during faults.
27. Transient stability can be enhanced by using a fast circuit breaker.
28. Grid frequency changes following a short circuit.
29. Transient stability is being simulated in computers using a modified Euler's method.
30. For transient stability analysis, one generator (aggregated) can be used instead of several parallel generators.

4.11 ANSWERS TO TWO-CHOICE QUESTIONS

2,5,6,9,10,12,14,16,18,19,23,26,29	No
1,3,4,7,8,11,13,15,17,20,21,22,24,25,27,28,30	Yes

4.12 DESCRIPTIVE QUESTIONS OF TRANSIENT STABILITY ANALYSIS

4.1 If the mechanical power changes suddenly, find the maximum angle.

$$P_{\text{max}} = 2 \text{ pu}, \ P_{m1} = 1.0 \text{ pu}, \ P_{m0} = 0.5 \text{ pu} \Rightarrow \delta_{\text{max}} = ?$$

Difficulty level • Easy o Normal o Hard

4.2 Demonstrate that two parallel generators behave similarly to an equivalent generator.

Difficulty level • Easy o Normal o Hard

If:

$$\frac{H_i}{\pi f^0}\frac{d^2\delta_i}{dt^2} = P_{mi} - P_{ei}, \quad i = 1,2$$

Prove:

$$\frac{H}{\pi f^0}\frac{d^2\delta}{dt^2} = P_m - P_e$$

That:

$$\delta = \delta_1 - \delta_2, \quad H = \frac{H_1 H_2}{H_1 + H_2}, \quad P_m = \frac{H_2 P_{m1} - H_1 P_{m2}}{H_1 + H_2}, \quad P_e = \frac{H_2 P_{e1} - H_1 P_{e2}}{H_1 + H_2}$$

4.3 Show that when the braking power is equal to one pu, the generator's speed drops to zero from its nominal value in $2H$ seconds using the swing equation.
   Difficulty level • Easy o Normal o Hard

4.4 Assume that: pre-fault system ($P_{max}\sin(\delta)$), faulted system ($r_1 P_{max}\sin(\delta)$) and post-fault system ($r_2 P_{max}\sin(\delta)$). Prove that the critical clearing angle is calculated from the following formula.
   Difficulty level • Easy o Normal o Hard

$$\cos\delta_{Cr} = \frac{\dfrac{P_m}{P_{max}}(\delta_{max} - \delta_0) + r_2 \cos\delta_{max} - r_1 \cos\delta_0}{r_2 - r_1}$$

4.5 Assume that 50% of the maximum power is the transmission power of the pre-fault system. The system experiences a short circuit, causing the reactance between the infinite bus and the generator to increase to 400% of its initial value. Seventy-five percent of the pre-fault maximum power is transmitted when the switch is closed, eliminating the short circuit. Find the critical clearing angle.
   Difficulty level • Easy o Normal o Hard

4.6 Consider a synchronous generator with the following specifications.

50 Hz, 2 – Pole, 20 MVA, 13.2 kV, $H = 9$ s, $P_{in} = 26,800$ hp, $P_{out} = 16$ MW

   a. Determine the stored energy at synchronous speed as well as the rotor acceleration.
   b. Determine the difference in the machine speed after 15 cycles, assuming the acceleration mentioned above is constant.
      Difficulty level o Easy • Normal o Hard

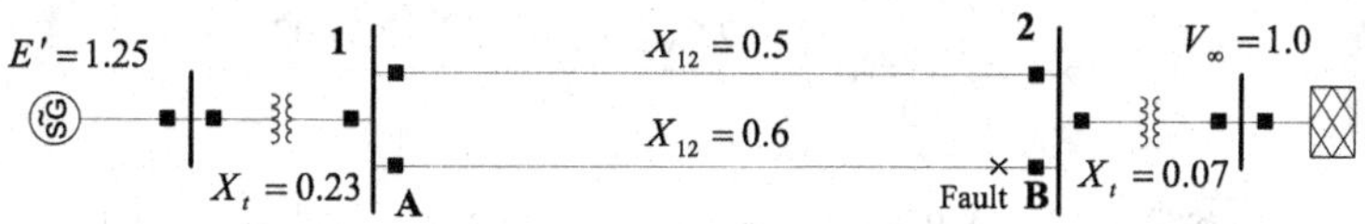

FIGURE 4.10 Question network 4.7.

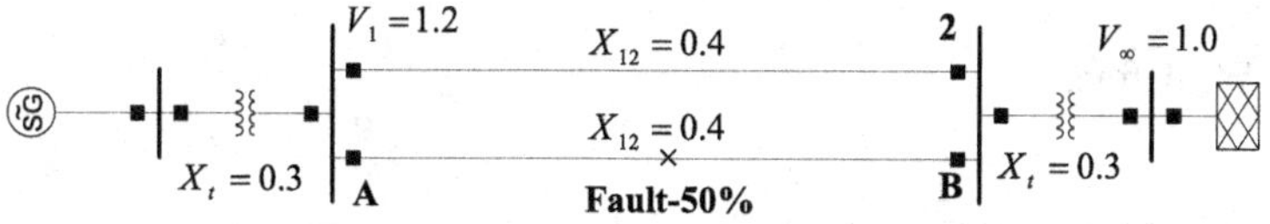

FIGURE 4.11 Question network 4.8.

4.7 (See Figure 4.10). Assume that $P_m = 1$ pu, in a post-fault system, both circuit breakers A and B simultaneously open the line. Find the critical clearing angle. A fault occurs near bus 2.
Difficulty level o Easy • Normal o Hard

4.8 (See Figure 4.11) Assume that $P_m = 1$ pu. In the middle of the AB line, there is a three-phase short circuit. At $60°$, the A and B circuit breakers will open simultaneously, removing the short circuit. Is the power system transiently stable?
Difficulty level o Easy o Normal • Hard

4.9 In question 4.8, what is (a) δ_{max} and (b) δ_{min}.
Assume that:

$$P_{e1}(\delta) = 1.746\,Sin\,\delta,\ P_{e2}(\delta) = 0.349\,Sin\,\delta,\ P_{e3}(\delta) = 1.397\,Sin\,\delta$$

Difficulty level o Easy o Normal • Hard

4.10 Consider a synchronous generator with the following specifications.

$$50\,Hz,\ H = 5\,s,\ P_m = 1\ pu,\ t_{cl} = 8\ cycle$$

$$Max(P_{e1}) = 1.8\ pu,\ Max(P_{e2}) = 0.6\ pu,\ Max(P_{e3}) = 1.35\ pu$$

Find δ_{cr}, δ_{cl}, and t_{cl}.
Difficulty level o Easy o Normal • Hard

4.13 DESCRIPTIVE ANSWERS OF TRANSIENT STABILITY ANALYSIS

4.1 From equation 4.35: $\delta_0 = Sin^{-1}\left(\dfrac{0.5}{2}\right) = 0.2527$ rad

$$\delta_{max} \triangleq \delta \Rightarrow 2\,Cos(\delta) + (\delta) + (-0.2527 - 2\,Cos(0.2527)) = 0 \Rightarrow$$

$$\Rightarrow 2\,Cos(\delta) + (\delta) - 2.1892 = 0$$

$$\Rightarrow \delta^{(r+1)} = \delta^{(r)} - \frac{2\cos\delta^{(r)} + \delta^{(r)} - 2.1892}{-2\sin\delta^{(r)} + 1}$$

$$\Rightarrow \delta^{(0)} = \frac{\pi}{2} \rightarrow \delta^{(1)} = 0.9524 \rightarrow \delta^{(2)} = 0.8296 \rightarrow \delta^{(3)} = 0.8101$$

$$\rightarrow \delta^{(4)} = 0.8095 \rightarrow \delta^{(5)} = \delta_{\max} = 0.8095 \text{ rad}$$

4.2   We define:

$$\frac{d^2\delta}{dt^2} = \frac{d^2}{dt^2}(\delta_1 - \delta_2) = \frac{d^2\delta_1}{dt^2} - \frac{d^2\delta_2}{dt^2} =$$

$$= \frac{\pi f^0}{H_1}(P_{m1} - P_{e1}) - \frac{\pi f^0}{H_2}(P_{m2} - P_{e2}) = \pi f^0\left(\frac{P_{m1}}{H_1} - \frac{P_{m2}}{H_2}\right) - \pi f^0\left(\frac{P_{e1}}{H_1} - \frac{P_{e2}}{H_2}\right) =$$

$$= \frac{\pi f^0}{H_1 H_2}(H_2 P_{m1} - H_1 P_{m2}) - \frac{\pi f^0}{H_1 H_2}(H_2 P_{e1} - H_1 P_{e2}) \qquad (4.41)$$

$$(\text{eq. } 4.41)\left(\frac{H_1 + H_2}{H_1 + H_2}\right) \Rightarrow$$

$$\Rightarrow \frac{d^2\delta}{dt^2} = \frac{\pi f^0}{\dfrac{H_1 H_2}{H_1 + H_2}}\left(\left(\frac{H_2 P_{m1} - H_1 P_{m2}}{H_1 + H_2}\right) - \left(\frac{H_2 P_{e1} - H_1 P_{e2}}{H_1 + H_2}\right)\right) = \frac{\pi f^0}{H}(P_m - P_e)$$

4.3   From the swing equation:

$$P_T - P_G = -1 \Rightarrow -1 = \frac{H}{\pi f^0}\frac{d^2\delta}{dt^2} \Rightarrow \frac{d\delta}{dt} = \frac{-\pi f^0}{H}t$$

$$\frac{d\delta}{dt} = \frac{-\pi f^0}{H}t \Rightarrow \omega - \omega_0 = -\frac{\pi f^0}{H}t \xrightarrow[\omega_0 = 2\pi f^0]{\omega = 0} t = 2H \text{ s}$$

4.4   From equation 4.32:

$$\delta_{cr} = \cos^{-1}\left(\frac{P_2 \cos\delta_0 - P_3 \cos\delta_{\max} - P_m(\delta_{\max} - \delta_0)}{(P_2 - P_3)}\right)$$

We have:

$$\Rightarrow \delta_{cr} = \cos^{-1}\left(\frac{r_1 P_{\max}\cos\delta_0 - r_2 P_{\max}\cos\delta_{\max} - P_m(\delta_{\max} - \delta_0)}{(r_1 P_{\max} - r_2 P_{\max})}\right)$$

$$\Rightarrow \operatorname{Cos}\delta_{Cr} = \frac{\dfrac{P_m}{P_{max}}(\delta_{max}-\delta_0)+r_2\operatorname{Cos}\delta_{max}-r_1\operatorname{Cos}\delta_0}{r_2-r_1}$$

4.5 See answer 4.4.

In pre-fault, we have:

$$\left\{\begin{array}{l} P_{max}=100\% \\ P_m=50\% \end{array}\right. \Rightarrow \left\{\begin{array}{l} P_{max}=1 \\ P_m=0.5 \end{array}\right.$$

In the faulted system, we know:

$$P\propto\frac{1}{X}\Rightarrow(X\uparrow\Rightarrow P\downarrow)\Rightarrow\frac{P_2}{P_1}=\frac{X_1}{X_2}=\frac{1}{400\%}=0.25=r_1$$

In the post-fault system, we have: $r_2=0.75$

$$\delta_0: P_m=P_{max}\operatorname{Sin}\delta\Rightarrow0.5=\operatorname{Sin}\delta\Rightarrow\delta_0=\frac{\pi}{6}=0.524\text{ rad}$$

$$\delta_{max}: P_m=r_2 P_{max}\operatorname{Sin}\delta\Rightarrow0.5=0.75\operatorname{Sin}\delta\Rightarrow$$

$$\delta_{max}=\pi-\operatorname{Sin}^{-1}\left(\frac{0.5}{0.75}\right)=2.412\text{ rad}$$

$$\Rightarrow \operatorname{Cos}\delta_{Cr}=\frac{\dfrac{0.5}{1}(2.412-0.524)+0.75\operatorname{Cos}2.412-0.25\operatorname{Cos}0.524}{0.75-0.25}=0.337$$

$$\Rightarrow \delta_{cr}=1.227\text{ rad}=70.30°$$

4.6 Assume that: $S_{base}=20$ MVA.

$$P_m=P_{in}=\frac{26800\times746}{20\times10^6}=1\text{ pu MW},\ P_{out}=P_e=\frac{16}{20}=0.8\text{ pu}$$

a. From the swing equation 4.2, the rotor acceleration is:

$$P_m-P_e=\frac{H}{\pi f^0}\frac{d^2\delta}{dt^2}\Rightarrow1-0.8=\frac{9}{\pi\times50}\frac{d^2\delta}{dt^2}\Rightarrow\frac{d^2\delta}{dt^2}=3.491\ \frac{\text{rad}}{\text{s}^2}$$

The stored energy at synchronous speed is:

$$W_{Kin}=HS_n=(9)(20)=180\text{ MJ}$$

b.   We have from (a):

$$\frac{d\delta}{dt} = 3.491t + \omega_0 \Rightarrow \delta = 1.746t^2 + \omega_0 t + \delta_0$$

We have:

$$\omega_0 = \frac{4\pi f_s}{p} = \frac{4\pi \times 50}{2} = 100\pi \ \frac{rad}{s}$$

The machine speed is:

$$t = \frac{15}{50} = 0.3\,\text{s} \Rightarrow \frac{d\delta}{dt} = 3.491 \times 0.3 + 100\pi = 315.21 \ \frac{rad}{s}$$

$$\Rightarrow \frac{d\delta}{dt} = 315.21 \frac{30}{\pi} = 3{,}010 \ \text{rpm}$$

4.7   In a pre-fault system, the total reactance is:

$$X_{tot} = 0.23 + (0.5 \parallel 0.6) + 0.07 = 0.573 \ \text{pu}$$

$$\Rightarrow P_{e1}(\delta) = \frac{E'V}{X_{tot}} \text{Sin}\,\delta = \frac{(1.25)(1)}{0.573} \text{Sin}\,\delta = 2.182 \ \text{Sin}\,\delta$$

In a faulted system, when a fault occurs on bus 2, no power is transmitted, then:

$$P_{e2}(\delta) = 0$$

In post-fault system, total reactance is:

$$X_{tot} = 0.23 + 0.5 + 0.07 = 0.8 \ \text{pu}$$

$$\Rightarrow P_{e3}(\delta) = \frac{(1.25)(1)}{0.8} \text{Sin}\,\delta = 1.563 \ \text{Sin}\,\delta$$

From equations 4.29–4.31:

$$\delta_0 : P_m = P_{e1}(\delta_0) \Rightarrow 1 = 2.182 \ \text{Sin}\,\delta_0 \Rightarrow \delta_0 = 0.476 \ \text{rad}$$

$$\delta_{max} : P_m = P_{e3}(\delta_{max}) \Rightarrow 1 = 1.563 \ \text{Sin}\,\delta_{max} \Rightarrow \delta_{max} = \pi - \text{Sin}^{-1}\left(\frac{1}{1.563}\right) = 2.447 \ \text{rad}$$

From equation 4.32:

$$\delta_{cr} = \text{Cos}^{-1}\left[\frac{P_m(\delta_{max} - \delta_0) + P_3\text{Cos}\,\delta_{max} - P_2\text{Cos}\,\delta_0}{P_3 - P_2}\right]$$

$$\Rightarrow \delta_{cr} = \text{Cos}^{-1}\left[\frac{(1)(2.447 - 0.476) + 1.563\text{Cos}\,2.447 - 0}{1.563 - 0}\right] = 1.056\,\text{rad} = 60.48°$$

4.8 Following is the formula for calculating the transmitted power between bus 1 and an infinite bus.

$$P_e = \frac{V_1 V_\infty}{X}\text{Sin}\,\delta_{V_1} \Rightarrow 1 = \frac{(1.2)(1)}{(0.4 \| 0.4) + 0.3}\text{Sin}\,\delta_{V_1} \Rightarrow \delta_{V_1} = \text{Sin}^{-1}\left(\frac{0.5}{1.2}\right) = 24.62°$$

Then:

$$\hat{I} = \frac{\hat{V}_1 - \hat{V}_\infty}{jX} = \frac{1.2\angle 24.62° - 1\angle 0}{j0.5} = 1.016\angle -10.31°$$

KVL:

$$\hat{E}' = (j0.3)\hat{I} + \hat{V}_1 = (j0.3)(1.016\angle -10.31°) + 1.2\angle 24.62° = 1.397\angle 34.92°$$

Then, in the pre-fault system, the total reactance and power are:

$$X_{tot} = 0.3 + (0.4 \| 0.4) + 0.3 = 0.8\,\text{pu}$$

$$P_{e1}(\delta) = \frac{E\,V_\infty}{X_{tot}}\text{Sin}\,\delta = \frac{1.397(1)}{0.8}\text{Sin}\,\delta = 1.746\,\text{Sin}\,\delta$$

Then, the equivalent circuit of the faulty power system is shown in Figure 4.12a. By converting the Δ into a Y, we get from Figure (a) to Figure (b), and by converting the Y into a Δ, we get from Figure (b) to Figure (c). From Figure 4.12.c:

$$P_{e2}(\delta) = \frac{1.397(1)}{4}\text{Sin}\,\delta = 0.349\,\text{Sin}\,\delta$$

Then, in a post-fault system, total reactance and power are:

$$X_{tot} = 0.3 + 0.4 + 0.3 = 1 \Rightarrow P_{e3}(\delta) = \frac{1.397(1)}{1}\text{Sin}\,\delta = 1.397\,\text{Sin}\,\delta$$

From equations 4.29–4.31:

$$\delta_0 : P_m = P_{e1}(\delta_0) \Rightarrow 1 = 1.746\,\text{Sin}\,\delta_0 \Rightarrow \delta_0 = 0.610\,\text{rad}$$

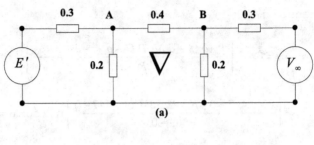

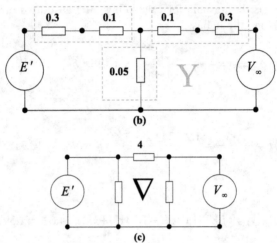

FIGURE 4.12   (a,b,c). Answer network 4.8.

$$\delta_1 : P_m = P_{e3}(\delta_1) \Rightarrow 1 = 1.397 \ \text{Sin} \, \delta_1 \Rightarrow \delta_1 = \text{Sin}^{-1}\left(\frac{1}{1.397}\right) = 0.798 \ \text{rad}$$

$$\delta_{\max} : P_m = P_{e3}\left(\delta_{\max}\right) \Rightarrow 1 = 1.397 \ \text{Sin} \, \delta_{\max} \Rightarrow \delta_{\max} = \pi - \delta_1 = 2.344 \ \text{rad}$$

From equation 4.32:

$$\Rightarrow \delta_{cr} = \text{Cos}^{-1}\left[\frac{(1)(2.344 - 0.610) + 1.397 \, \text{Cos} \, 2.344 - 0.349 \, \text{Cos} \, 0.61}{1.397 - 0.349}\right] \Rightarrow$$

$$\delta_{cr} = 1.103 \ \text{rad} = 63.22° > 60°$$

Because ($\delta_{cr} > 60$), then the system is stable.

4.9   We have:

a.

$$P_{e1}(\delta) = 1.746 \ \text{Sin} \, \delta, \ P_{e2}(\delta) = 0.349 \ \text{Sin} \, \delta, \ P_{e3}(\delta) = 1.397 \ \text{Sin} \, \delta, \ P_m = 1$$

$$\delta_0 = 0.610 \ \text{rad}, \ \delta_{cl} = 60° = \frac{\pi}{3} = 1.047 \ \text{rad}, \ \delta_1 = 0.798 \ \text{rad}$$

From equation 4.22, the acceleration area is:

$$A_{acc} = \int_{\delta_0}^{\delta_{cl}} \left(P_m - P_{e_2}(\delta) \right) d\delta = \int_{0.610}^{1.047} \left(1 - 0.349 \operatorname{Sin} \delta \right) d\delta = 0.326$$

From equation 4.23, the deceleration area is:

$$A_{dec} = \int_{1.047}^{\delta_{max}} \left(1.397 \operatorname{Sin} \delta - 1 \right) d\delta = -1.397 \operatorname{Cos} \delta - \delta \Big|_{1.047}^{\delta_{max}} \Rightarrow$$

$$A_{dec} = \left(-1.397 \operatorname{Cos} \delta_{max} - \delta_{max} \right) - \left(-1.397 \operatorname{Cos} 1.047 - 1.047 \right) \Rightarrow$$

$$A_{dec} = -1.397 \operatorname{Cos} \delta_{max} - \delta_{max} + 1.746 \Rightarrow$$

From equation 4.24, stability condition:

$$A_{acc} = A_{dec} \Rightarrow 0.326 = -1.397 \operatorname{Cos} \delta_{max} - \delta_{max} + 1.746 \Rightarrow$$

$$-1.397 \operatorname{Cos} \delta_{max} - \delta_{max} + 1.420 = 0 \Rightarrow$$

From equation 4.36, iterative methods:

$$\delta_{max}^{(r+1)} = \delta_{max}^{(r)} - \frac{-1.397 \operatorname{Cos} \delta_{max}^{(r)} - \delta_{max}^{(r)} + 1.420}{1.397 \operatorname{Sin} \delta_{max}^{(r)} - 1}$$

We guess:

$$r = 0, \ \delta_{max}^{(0)} = \frac{\pi}{2} = 1.571 \Rightarrow$$

$$\delta_{max}^{(1)} = 1.571 - \frac{-1.397 \operatorname{Cos}(1.571) - 1.571 + 1.420}{1.397 \operatorname{Sin}(1.571) - 1} = 1.951 \text{ rad}$$

$$\delta_{max}^{(2)} = 1.951 - \frac{-1.397 \operatorname{Cos}(1.951) - 1.951 + 1.420}{1.397 \operatorname{Sin}(1.951) - 1} = 1.993 \text{ rad}$$

$$\delta_{max}^{(3)} = 1.993 - \frac{-1.397 \operatorname{Cos}(1.993) - 1.993 + 1.420}{1.397 \operatorname{Sin}(1.993) - 1} = 1.995 \text{ rad}$$

$$\delta_{max}^{(4)} = 1.995 - \frac{-1.397 \operatorname{Cos}(1.995) - 1.995 + 1.420}{1.397 \operatorname{Sin}(1.995) - 1} = 1.995 \text{ rad}$$

$$\delta_{max}^{(4)} = \delta_{max}^{(3)} \Rightarrow \delta_{max} = 1.995 \text{ rad} = 114.3°$$

Now we check the deceleration area:

$$A_{\text{dec}} = \int_{1.047}^{1.995} \left(1.397 \operatorname{Sin} \delta - 1\right) d\delta = 0.326 \Rightarrow A_{\text{dec}} = A_{\text{acc}}$$

b.

From equation 4.25, the acceleration area is:

$$A_{\text{acc}} = \int_{\delta_1}^{\delta_{\max}} \left(P_{e_3}(\delta) - P_m\right) d\delta = \int_{0.798}^{1.995} \left(1.397 \operatorname{Sin} \delta - 1\right) d\delta = 0.353$$

From equation 4.26, the deceleration area is:

$$A_{\text{dec}} = \int_{\delta_{\min}}^{\delta_1} \left(P_m - P_{e_3}(\delta)\right) d\delta = \int_{\delta_{\min}}^{0.798} \left(1 - 1.397 \operatorname{Sin} \delta\right) d\delta$$

$$A_{\text{dec}} = \delta + 1.397 \operatorname{Cos} \delta \big|_{\delta_{\min}}^{0.798} = (0.798 + 1.397 \operatorname{Cos} 0.798) - \left(\delta_{\min} + 1.397 \operatorname{Cos} \delta_{\min}\right)$$

$$\Rightarrow A_{\text{dec}} = 1.773 - \delta_{\min} - 1.397 \operatorname{Cos} \delta_{\min}$$

From equation 4.27, stability condition:

$$A_{\text{dec}} = A_{\text{acc}} \Rightarrow 0.353 = 1.773 - \delta_{\min} - 1.397 \operatorname{Cos} \delta_{\min}$$

$$1.420 - \delta_{\min} - 1.397 \operatorname{Cos} \delta_{\min} = 0$$

From equation 4.36, iterative methods:

$$\delta_{\min}^{(r+1)} = \delta_{\min}^{(r)} - \frac{-1.397 \operatorname{Cos} \delta_{\min}^{(r)} - \delta_{\min}^{(r)} + 1.420}{1.397 \operatorname{Sin} \delta_{\min}^{(r)} - 1}$$

We guess:

$$r = 0,\ \delta_{\min}^{(0)} = 0 \Rightarrow$$

$$\delta_{\min}^{(1)} = 0 - \frac{-1.397 \operatorname{Cos}(0) - 0 + 1.420}{1.397 \operatorname{Sin}(0) - 1} = 0.023$$

$$\delta_{\min}^{(2)} = 0.023 - \frac{-1.397 \operatorname{Cos}(0.023) - 0.023 + 1.420}{1.397 \operatorname{Sin}(0.023) - 1} = 0.023$$

$$\delta_{\min}^{(2)} = \delta_{\min}^{(1)} \Rightarrow \delta_{\min} = 0.023 \,\text{rad} = 1.34°$$

Now we check the deceleration area:

$$A_{dec} = \int_{0.023}^{0.798} \left(1 - 1.397\,\text{Sin}\,\delta\right)\,d\delta = 0.354 \approx 0.353 = A_{acc}$$

4.10 We have from equations 4.29–4.31:

$$\delta_0 : P_m = P_{e1}(\delta_0) \Rightarrow 1 = 1.8\,\text{Sin}\,\delta_0 \Rightarrow \delta_0 = 0.589\ \text{rad}$$

$$\delta_1 : P_m = P_{e3}(\delta_1) \Rightarrow 1 = 1.35\ \text{Sin}\,\delta_1 \Rightarrow \delta_1 = 0.834\ \text{rad}$$

$$\delta_{max} = \pi - 0.834 = 2.307\,\text{rad} \Rightarrow$$

From equation 4.32:

$$\Rightarrow \delta_{cr} = \text{Cos}^{-1}\left[\frac{(1)(2.307 - 0.589) + 1.35\,\text{Cos}\,2.307 - 0.6\,\text{Cos}\,0.589}{1.35 - 0.6}\right] = 65.36°$$

Now the power swing equation for the faulted system is:

$$\frac{d^2\delta}{dt^2} = \frac{\pi f^0}{H}\left(1 - P_{e2}(\delta)\right) = \frac{50\pi}{5}\left(1 - 0.6\,\text{Sin}(\delta)\right) = 31.416 - 18.85\,\text{Sin}(\delta)$$

From equation 4.39, modified Euler's method:

$$X_1 = \delta,\ X_2 = \omega = \dot{\delta}$$

We solved the time of eight cycles in **three steps**.

$$\Delta t = 8 \times \frac{1}{50} \times \frac{1}{3} = 0.053\ \text{s}$$

$$\Rightarrow \begin{cases} \dot{\delta} = \omega \\ \dot{\omega} = 31.42 - 18.85\,\text{Sin}\,\delta \end{cases}$$

$$C : \begin{cases} \delta_{i+1}^c = \delta_i + \omega_i \Delta t \\ \omega_{i+1}^c = \omega_i + (31.42 - 18.85\,\text{Sin}\,\delta_i)\Delta t \end{cases}$$

$$P : \begin{cases} \delta_{i+1}^p = \delta_i + 0.5(\omega_i + \omega_{i+1}^c)\Delta t \\ \omega_{i+1}^p = \omega_i + 0.5(31.42 - 18.85\,\text{Sin}\,\delta_i + 31.42 - 18.85\,\text{Sin}\,\delta_{i+1}^c)\Delta t \end{cases}$$

In step 1 ($i = 0$) we have:

$$\begin{cases} \delta_0 = 0.589 \text{ rad} \\ \omega_0 = 0 \end{cases}$$

$$C : \begin{cases} \delta_1^c = \delta_0 + \omega_0 \Delta t = 0.589 + 0 = 0.589 \\ \omega_1^c = \omega_0 + (31.42 - 18.85 \sin \delta_0) \Delta t = \\ = 0 + (31.42 - 18.85 \sin 0.589)0.053 = 1.110 \end{cases}$$

$$P : \begin{cases} \delta_1^p = \delta_1 = \delta_0 + 0.5(\omega_0 + \omega_1^c)\Delta t = 0.589 + 0.5(0 + 1.110)0.053 = 0.618 \\ \omega_1^p = \omega_1 = \omega_0 + 0.5(2(31.42) - 18.85 \sin \delta_0 - 18.85 \sin \delta_1^c)\Delta t = \\ = 0 + 0.5(62.84 - 18.85 \sin 0.589 - 18.85 \sin 0.589)0.053 = 1.110 \end{cases}$$

In step 2 ($i = 1$) we have:

$$C : \begin{cases} \delta_2^c = \delta_1 + \omega_1 \Delta t = 0.618 + (1.110)0.053 = 0.677 \\ \omega_2^c = \omega_1 + (31.42 - 18.85 \sin \delta_1) \Delta t = \\ = 1.110 + (31.42 - 18.85 \sin 0.618)0.053 = 2.196 \end{cases}$$

$$P : \begin{cases} \delta_2^p = \delta_2 = \delta_1 + 0.5(\omega_1 + \omega_2^c)\Delta t = 0.618 + 0.5(1.110 + 2.196)0.053 = 0.706 \\ \omega_2^p = \omega_2 = \omega_1 + 0.5(2(31.42) - 18.85 \sin \delta_1 - 18.85 \sin \delta_2^c)\Delta t \\ = 1.110 + 0.5(62.84 - 18.85 \sin 0.618 - 18.85 \sin 0.677)0.053 = \\ = 2.173 \end{cases}$$

In step 3 ($i = 2$) we have:

$$C : \begin{cases} \delta_3^c = \delta_2 + \omega_2 \Delta t = 0.706 + (2.173)0.053 = 0.821 \\ \omega_3^c = \omega_2 + (31.42 - 18.85 \sin \delta_2) \Delta t = \\ = 2.173 + (31.42 - 18.85 \sin 0.706)0.053 = 3.190 \end{cases}$$

$$P : \begin{cases} \delta_3^p = \delta_3 = \delta_2 + 0.5(\omega_2 + \omega_3^c)\Delta t = 0.706 + 0.5(2.173 + 3.190)0.053 = 0.848 \\ \omega_3^p = \omega_3 = \omega_2 + 0.5(2(31.42) - 18.85 \sin \delta_2 - 18.85 \sin \delta_3^c)\Delta t \\ = 2.173 + 0.5(62.84 - 18.85 \sin 0.706 - 18.85 \sin 0.821)0.053 = 3.149 \end{cases}$$

$$\Rightarrow \delta_3 = \delta_{cl} = 0.848 \times \frac{180}{\pi} = 48.59°$$

So:

$$(\delta_{cl} = 48.59) < (\delta_{cr} = 65.36)$$

Then, the system is stable.

5 Power System Linear Controls

Part One: Lesson Summary

5.1 INTRODUCTION

Automatic control systems play a crucial role in power systems for maintaining stability, efficiency, and reliability. There are two types of control; **local controls** which are employed at turbine-generator units, and at selected voltage-controlled buses, and **central controls** which are employed at area control centers.

This section briefly reviews power system linear controls. In this case, a little load changes cause small changes in the entire network, especially in the voltages and frequency. In this chapter, power system linear controllers are primarily discussed in terms of their steady-state response. In the upcoming sections, the focus will be on discussing two separate control loops. Following a load change, the automatic voltage regulator (AVR) maintains the generator voltage. ALFC (automatic load frequency control) maintains the generator and turbine's frequency and speed.

By engaging with the two-choice questions and tackling them, the reader will enhance their comprehension of this chapter.

5.2 GENERATOR CONTROL LOOPS

(See Figure 5.1) Two control loops are present in synchronous generators: automatic voltage regulator (AVR) loop and automatic load frequency control (ALFC) loop. Because there is a mechanical steam valve involved in the ALFC loop, its response speed is notably slower compared to the speed of the AVR loop.

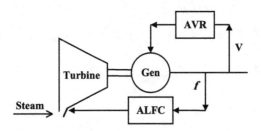

FIGURE 5.1 Synchronous generator AVR and ALFC loops.

DOI: 10.1201/9781003506751-5

5.3 AUTOMATIC VOLTAGE REGULATOR (AVR) LOOP

Figure 5.2 shows a simplified model of an AVR. For the initial analysis, it is possible to ignore the stabilizer, rectifier, and saturation model. For each element, its control block should be obtained which we ignore.

Figure 5.3 shows the AVR control blocks model. These control blocks can be mathematically expressed by equations 5.1–5.5.

We define:

$$G(s) \triangleq G_A(s) G_E(s) G_g(s) \qquad (5.1)$$

In the absence of any other signals, we obtain:

$$\frac{\Delta V(s)}{\Delta V_{ref}(s)} = \frac{G(s)}{1 + G(s)} \qquad (5.2)$$

Now the steady-state response can be obtained for a step input as:

$$\Delta V_{ref}(s) = \frac{\Delta V_{ref}}{s}, \ \Delta V_{ss} = \Delta V_t(t \to \infty) = \lim_{s \to 0}(s\,\Delta V(s)) \qquad (5.3)$$

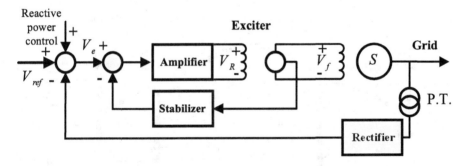

FIGURE 5.2 A simplified AVR model.

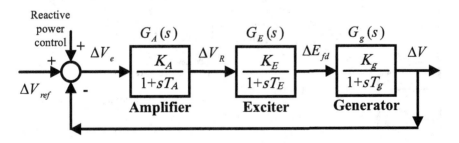

FIGURE 5.3 The AVR control blocks model.

From equations 5.2 and 5.3:

$$\Delta V_{ss} = \lim_{s \to 0} \left(s \, \frac{\Delta V_{ref}}{s} \, \frac{G(s)}{1+G(s)} \right) = \Delta V_{ref} \, \frac{G(0)}{1+G(0)} \tag{5.4}$$

From equation 5.1:

$$K \triangleq G(0) = K_A \, K_E \, K_g \Rightarrow \Delta V_{ss} = \Delta V_{ref} \, \frac{K}{1+K} \tag{5.5}$$

5.4   AUTOMATIC LOAD FREQUENCY CONTROL (ALFC) LOOP

Figure 5.4 shows an ALFC simplified model for a steam power plant[1]. It is important to keep in mind the next two points to evaluate ALFC's performance. The bar (AC) does not bend from the linkage point (B). Also, the bar (CE) does not bend from the linkage point (D).

Now we examine the next two scenarios.

1. Scenario 1: The steam valve needs to be opened if the load increases.
2. Scenario 2: The steam valve needs to be closed if a command to the frequency reduction is issued.

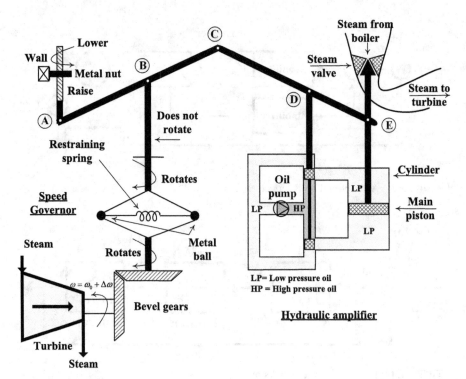

FIGURE 5.4   An ALFC simplified model.

5.4.1   SCENARIO 1

Follow Figure 5.5 with these steps:

1. In this scenario, linkage point (A) is fixed (0). [There has been no central control command issued.]
2. Load increase (↑) → Reducing network energy (↓) → Reducing kinetic energy (↓) → Reducing the speed of rotating machines (↓) → Reducing turbine speed (↓).
3. In defiance of the law of centrifugation, spring force brings two flyweights together (→←).
4. Linkage point (B) moves up (↑) (linkage point A is fixed).
5. Linkage point (C) also moves up (↑).
6. In this case, linkage point (E) is fixed (0) due to the fixedness of two low-pressure oils (LP1 = LP2) in the cylinder.
7. When (C) moves up (↑), linkage point (D) also moves up (↑).
8. High-pressure oil (HP) moves to the upper part of the cylinder.
9. Linkage point (E) and the steam valve pull down (↓) (steam valve opens).
10. In this case, linkage point (C) is fixed (0) (because points A and B are fixed).

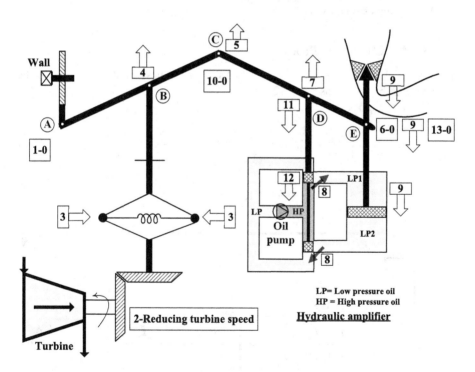

FIGURE 5.5   Steam valve opening with load reduction.

11. Therefore, when linkage point (E) moves down ($\downarrow$), linkage point (D) also moves down ($\downarrow$).
12. Then the movement of high-pressure oil (HP) to the upper part of the cylinder stops (0).
13. The downward movement of linkage point (E) stops (0).
14. All points stop moving (0).

5.4.2   SCENARIO 2

We analyze this scenario more quickly.

1. Point (B) is fixed (0). (The frequency is constant.)
2. Point (A) moves up ($\uparrow$).
3. Point (C) moves down ($\downarrow$).
4. Point (E) is fixed (0).
5. Point (D) moves down ($\downarrow$).
6. High-pressure oil (HP) moves to the lower part of the cylinder.
7. Point (E) and the steam valve pull up ($\uparrow$) (steam valve closes).
8. In this case, point (C) is fixed (0).
9. Therefore, when point (E) moves up ($\uparrow$), point (D) also moves up ($\uparrow$).
10. All points stop moving (0).

5.4.3   ALFC CONTROL BLOCKS

Figure 5.6 shows the ALFC control blocks model. These control blocks can be summarized by equations 5.6–5.12.

The control blocks:

$$G_H(s) = \frac{k_H}{1+sT_H}, \ G_T(s) = \frac{k_T}{1+sT_T}, \ G_P(s) = \frac{k_P}{1+sT_P} = \frac{1}{D+s(2H)} \qquad (5.6)$$

$$k_P = \frac{1}{D}, \ T_P = \frac{2H}{D} \qquad (5.7)$$

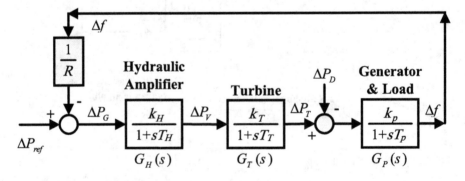

FIGURE 5.6   The ALFC control blocks model.

where:

R is the governor's speed regulation or **droop** in

$$\frac{\text{pu Hz}}{\text{pu MW}} \tag{5.8}$$

ΔP_D is a frequency-independent load in pu.
D is the load sensitivity coefficient to the frequency in

$$\frac{\text{pu MW}}{\text{pu Hz}} \tag{5.9}$$

H is the **generator inertia constant** in seconds and defines:

$$H = \frac{W_{\text{kin}}}{S_{\text{base}}} = \frac{\text{Kinetic energy}}{\text{Base power}} \left(\frac{\text{MJ}}{\text{MVA}} = \text{s} \right) \tag{5.10}$$

T_T, T_H, and T_P are turbine, hydraulic, and loaded generator **time constants**, respectively.

Note 1: From Figure 5.6 and equation 5.10, we have:

$$(\Delta P_T - \Delta P_D)\left(\frac{1}{D+s(2H)} \right) = \Delta f \Rightarrow (\Delta P_T - \Delta P_D) = (D+s(2H))\Delta f \Rightarrow$$

$$(\Delta P_T - \Delta P_D) = D\Delta f + (2H)\frac{d}{dt}\Delta f \Rightarrow \tag{5.11}$$

$$(\Delta P_T - \Delta P_D) = D\Delta f + 2\left(\frac{W_{\text{kin}}}{S_b} \right)\frac{d}{dt}\Delta f \tag{5.12}$$

Note 2: If the ΔP_{ref} is zero, the steady-state response can be obtained for the step input.

$$\Delta P_D(s) = \frac{\Delta P_D}{s}, \quad \Delta f_{ss} = \Delta \omega_{ss} = \Delta f(t \to \infty) = \lim_{s \to 0}(s\,\Delta f(s)) \tag{5.13}$$

Then assume $k_H = k_T = 1$:

$$\Delta f_{ss} = \Delta \omega_{ss} = \frac{G_P(0)(-\Delta P_D)}{1 + \dfrac{1}{R}G_H(0)G_T(0)G_P(0)} = \frac{-k_P}{1 + \dfrac{k_P}{R}}\Delta P_D = \frac{-\Delta P_D}{D + \dfrac{1}{R}} \tag{5.14}$$

This relationship is also valid for several parallel regions (if the bases are the same).

$$D = D_1 + D_2 + \cdots, \quad \frac{1}{R} = \frac{1}{R_1} + \frac{1}{R_2} + \cdots \tag{5.15}$$

Note 3: In this section and in the pu mode:

$$\omega \; \text{pu} = \frac{\omega \; \dfrac{\text{rad}}{\text{s}}}{\omega_b \; \dfrac{\text{rad}}{\text{s}}} = \frac{2\pi f \; \text{Hz}}{2\pi f_b \; \text{Hz}} = \frac{f \; \text{Hz}}{f_b \; \text{Hz}} = f \; \text{pu} \tag{5.16}$$

Note 4: The following happens to the new values of (R) and (D), if the base power is switched from $\left(S_b^{old}\right)$ to $\left(S_b^{new}\right)$.

$$R^{new} = R^{old}\left(\frac{S_b^{new}}{S_b^{old}}\right), \quad D^{new} = D^{old}\left(\frac{S_b^{old}}{S_b^{new}}\right) \tag{5.17}$$

Note 5: We obtain the crucial equation 5.18 from equations 5.14 and 5.15. The following modifications to its components are what cause the (ΔP_D).

$$\Delta f_{ss} = \frac{-\Delta P_D}{D_1 + D_2 + \cdots + \dfrac{1}{R_1} + \dfrac{1}{R_2} + \cdots} \Rightarrow$$

$$\Delta P_D = \left(-\Delta f_{ss} D_1\right) + \left(-\Delta f_{ss} D_2\right) + \cdots + \left(-\Delta f_{ss}\frac{1}{R_1}\right) + \left(-\Delta f_{ss}\frac{1}{R_2}\right) + \cdots \tag{5.18}$$

That:

Changing the frequency-dependent loads of the i th area:

$$-\Delta f_{ss} D_i \tag{5.19}$$

Changing the production of generators in the i th area:

$$\Delta P_{Ti} = -\Delta f_{ss}\frac{1}{R_i} \tag{5.20}$$

5.4.4 Using the Integral Block in ALFC

(See Figure 5.7). It is possible to remove the steady-state error by applying an integral block.

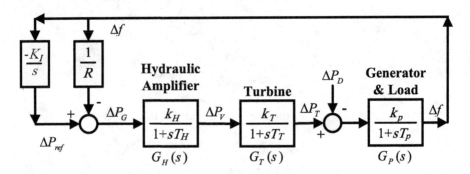

FIGURE 5.7 The ALFC control blocks model with integral block.

5.4.5 ALFC IN MULTI-GENERATORS NETWORK

This is the linearization of the transmission power equation between two networks connected by a transmission line.

$$P_{12} = \frac{V_1 V_2}{x_{12}} \operatorname{Sin}(\delta_{12}) \Rightarrow \frac{\Delta P_{12}}{\Delta \delta_{12}} \simeq \frac{dP_{12}}{d\delta_{12}}\bigg|_{\delta_{12}^0} = \frac{V_1 V_2}{x_{12}} \operatorname{Cos}(\delta_{12}^0) \triangleq T_{12}^0 \Rightarrow$$

$$\Delta P_{12} = T_{12}^0 (\Delta \delta_1 - \Delta \delta_2), \ \ \Delta \delta = \int \Delta \omega, \ \Rightarrow \Delta P_{12}(s) = T_{12}^0 \frac{\Delta \omega_1(s) - \Delta \omega_2(s)}{s} \Rightarrow$$

From equation 5.16:

$$\Delta P_{12}(s) = \frac{T_{12}^0}{s} \big(\Delta f_1(s) - \Delta f_2(s) \big) \tag{5.21}$$

T_{12}^0 is tie line synchronizing power coefficient (See Figure 5.8). More networks can be connected by extending equation 5.21.

The steady-state response can be obtained for a step input. We have:

$$\Delta f_{ss} = \Delta \omega_{ss} \text{ pu} = \frac{-\Delta P_{D1} - \cdots - \Delta P_{Dn}}{(D_1 + \cdots + D_n) + \left(\dfrac{1}{R_1} + \cdots + \dfrac{1}{R_n} \right)} = \frac{-\Delta P_D}{D + \dfrac{1}{R}} \tag{5.22}$$

where:

$$\Delta P_D = \Delta P_{D1} + \cdots + \Delta P_{Dn}, \ \ D = D_1 + \cdots + D_n, \ \ \frac{1}{R} = \frac{1}{R_1} + \cdots + \frac{1}{R_n} \tag{5.23}$$

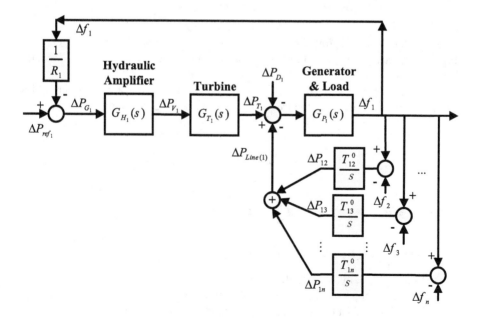

FIGURE 5.8 The ALFC control blocks model in a multi-generator network.

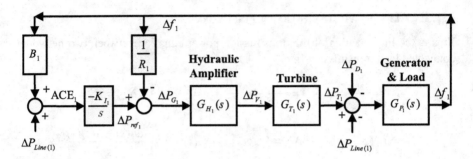

FIGURE 5.9 The ALFC control blocks model with integral blocks in a multi-generator network.

The same format can be used to write equation 5.22.

$$\Delta f_{ss} = \Delta \omega_{ss} \text{ pu} = \frac{-\Delta P_{D1} - \cdots - \Delta P_{Dn}}{\left(D_1 + \dfrac{1}{R_1}\right) + \cdots + \left(D_n + \dfrac{1}{R_n}\right)} = \frac{-\Delta P_D}{B_1 + \cdots + B_n} \tag{5.24}$$

where B denotes the **frequency bias factor** also known as the **actual frequency response characteristic** (AFRC). And:

$$\Delta P_{\text{Line}(i)} = \Delta P_{i1} + \cdots + \Delta P_{in} = -B_i \Delta f - \Delta P_{Di}, \; \Delta P_{ii=0} \tag{5.25}$$

For $n = 2$ and $i = 1$, we have:

$$\Delta P_{\text{Line}(1)} = \Delta P_{12} = -B_1 \Delta f - \Delta P_{D1} = \frac{B_1 \Delta P_{D2} - B_2 \Delta P_{D1}}{B_1 + B_2} \tag{5.26}$$

The problem section provides the proofs for the formulas in this chapter. (See Figure 5.9). It is possible to remove the steady-state error by applying the integral block.

The area control error (ACE) shows overproduction in the area. The area bias (B_i) in Figure 5.9 is equal to the frequency bias factor under ideal circumstances. We have:

$$B_i = D_i + \frac{1}{R_i} \tag{5.27}$$

Part Two: Answer-Question

5.5 TWO-CHOICE QUESTIONS (YES/NO)

1. In the AVR loop, a hydraulic amplifier is employed.
2. The LFC loop has a lower speed than the AVR loop.
3. The AVR and LFC loops are almost independent of each other.

4. An integrator provides stability to the AVR loop.
5. The steam valve will close if the turbine speed or frequency increases.
6. The steam valve will close if the load increases.
7. The frequency and speed of the turbine provide feedback to the loop (LFC).
8. In (LFC), the generator is represented by the superior pole.
9. A higher droop indicates a more frequency-dependent generator.
10. Frequency changes are improved by connecting two networks.
11. Having a frequency of 49.9 Hz in a network means reducing load.
12. If one generator's control loop opens, the frequency drops more.
13. Frequency drops will increase if the loads do not depend on frequency ($D=0$).
14. If the line between two grids is opened, the frequency drops more.
15. The integrator's input needs to be zero in the steady-state conditions.

5.6 ANSWERS TO TWO-CHOICE QUESTIONS

1,4,6,9,11	No
Other	Yes

5.7 DESCRIPTIVE QUESTIONS OF POWER SYSTEM LINEAR CONTROLS

5.1 If the steady-state error is less than 1% of the input value, find the loop gain in the AVR loop.
Difficulty level • Easy o Normal o Hard

5.2 An out-of-circuit generator accounts for 8% of the network's total capacity. Determine the frequency change rate in Hertz per second.

$$H=4\,\text{s}, f^0=50\,\text{Hz}$$

Difficulty level • Easy o Normal o Hard

5.3 The generator's power rises by 5 MW in a network. Determine the frequency change rate in Hz/s.

$$W_{\text{kin}}=1,500\,\text{MJ}, f^0=50\,\text{Hz}$$

Difficulty level • Easy o Normal o Hard

5.4 The load in a network is decreased by 40 MW. After 0.4 s, determine the frequency.

$$H=5\,\text{s}, f^0=50\,\text{Hz}, S_n=80\,\text{MW}$$

Difficulty level • Easy o Normal o Hard

5.5 How do the following two parallel generators share the 500 MW load?
Difficulty level • Easy o Normal o Hard

$$G_1: (S_{n1}=200 \text{ MW}, R_1=2\%), G_2: (S_{n2}=600 \text{ MW}, R_2=3\%)$$

5.6 Repeat question 5.5 for three generators.

$$G_3: (S_{n3}=250 \text{ MW}, R_3=5\%)$$

Difficulty level • Easy o Normal o Hard
5.7 Repeat question 5.5. How do the following two parallel generators share the 500 MW load?
Difficulty level o Easy • Normal o Hard

$$G_1: (S_{n1}=200 \text{ MW}, R_1=2\%, D_1=10 \text{ pu})$$

$$G_2: (S_{n2}=600 \text{ MW}, R_2=3\%, D_2=5 \text{ pu})$$

5.8 Two generators are connected by a line. The load in area 1, 500 MW is added. How is the load divided between the two generators? $(S_n=500)$

$$G_1: (R_1=2\%, D_1=10 \text{ pu}), G_2: (R_2=3\%, D_2=5 \text{ pu}), \text{ Line: } (T_{12}=3 \text{ pu})$$

Difficulty level o Easy o Normal • Hard
5.9 In question **5.8**, prove that if the line between two grids is opened, the frequency drops more.
Difficulty level o Easy • Normal o Hard
5.10 In question **5.5**, prove that if the first generator's control loop is opened, the frequency drops more.
Difficulty level o Easy • Normal o Hard
5.11 Prove that (Δf) and (ΔP_{line}) become zero when integrals are present in the two-area control loop.
Difficulty level o Easy • Normal o Hard
5.12 Two generators are connected by a line. $(S_b=500 \text{ MVA})$

$$\text{If } \Delta P_{D1}=500 \text{ MW then } \Delta P_{21}=195 \text{ MW}$$

$$\text{If } T_{line}=0 \text{ then } f_1(\infty)=59 \text{ Hz and } f_2(\infty)=60 \text{ Hz}$$

$$\text{If } D_1=10 \text{ and } D_2=5, \text{ find } R_1 \text{ and } R_2.$$

Difficulty level o Easy o Normal • Hard
5.13 The following parameters are assumed in the AVR loop.

$$T_A=0.1 \text{ s}, T_E=0.5 \text{ s}, T_g=5 \text{ s}$$

Using the root-locus diagram, find the stable gain (K).

$$K = K_A K_E K_g$$

5.8 DESCRIPTIVE ANSWERS OF POWER SYSTEM LINEAR CONTROL

5.1 From equation 5.5 and Figure 5.3, we have:

$$\Delta V_e = \Delta V_{ref} - \Delta V_{ss} = \Delta V_{ref} - \Delta V_{ref}\frac{K}{1+K} = \frac{\Delta V_{ref}}{1+K} < (1\%)\left(\Delta V_{ref}\right) \Rightarrow$$

$$1 + K > 100 \Rightarrow K > 99$$

5.2 From equation 5.6 and Figure 5.6, we have:
  a. The generator does not have output power, but it is in the network. From equation 5.11, we have:

$$\left(\Delta P_T - \Delta P_D\right) = D\Delta f + (2H)\frac{d}{dt}\Delta f$$

$$\left(0 - 8\%\right) = 0 + (2H)\frac{d}{dt}\Delta f \Rightarrow$$

$$-0.08 \text{ pu} = 2H\frac{d}{dt}\Delta f \Rightarrow$$

$$-\frac{0.08 \text{ pu}}{(2)(4\,\text{s})} = \frac{d}{dt}\Delta f \Rightarrow \frac{d}{dt}\Delta f = -0.01\frac{\text{pu}}{\text{s}} \Rightarrow$$

$$\Rightarrow \frac{d}{dt}\Delta f = -(0.01)(50 \text{ Hz}) = -0.5\frac{\text{Hz}}{\text{s}}$$

  b. The generator output power is zero and it is not in the network. From equation 5.12:

$$\left(\Delta P_T - \Delta P_D\right) = D\Delta f + 2\left(\frac{W_{kin}}{S_b}\right)\frac{d}{dt}\Delta f$$

$$\left(\Delta P_T - \Delta P_D\right) = D\Delta f + 2\left(\frac{W_{kin} - 0.08W_{kin}}{S_b}\right)\frac{d}{dt}\Delta f \Rightarrow$$

$$\left(\Delta P_T - \Delta P_D\right) = D\Delta f + (2)(0.92)\left(\frac{W_{kin}}{S_b}\right)\frac{d}{dt}\Delta f \Rightarrow$$

$$(\Delta P_T - \Delta P_D) = D\Delta f + (2)(0.92)(H)\frac{d}{dt}\Delta f \Rightarrow$$

$$0 - 0.08 = 0 + (2)(0.92)(4)\frac{d}{dt}\Delta f \Rightarrow$$

$$\frac{d}{dt}\Delta f = -10.870 \times 10^{-3} \frac{\text{pu}}{\text{s}}$$

$$\frac{d}{dt}\Delta f = (-10.870 \times 10^{-3})(50 \text{ Hz}) = -0.544 \frac{\text{Hz}}{\text{s}}$$

5.3 From equation 5.12, we have:

$$(\Delta P_T - \Delta P_D) = D\Delta f + \left(2\frac{W_{\text{kin}}}{S_b}\right)\frac{d}{dt}\Delta f$$

$$\left(\frac{5\,\text{MW}}{S_b\,\text{MVA}} - 0\right) = 0 + (2)\left(\frac{1,500\,\text{MJ}}{S_b\,\text{MVA}}\right)\frac{d}{dt}\Delta f \Rightarrow$$

$$\frac{d}{dt}\Delta f = \frac{5}{(2)(1,500)} = 1.667 \times 10^{-3} \frac{\text{pu}}{\text{s}}$$

$$\frac{d}{dt}\Delta f = (1.667 \times 10^{-3})(50 \text{ Hz}) = 0.083 \frac{\text{Hz}}{\text{s}}$$

5.4 From equation 5.11, we have:

$$(\Delta P_T - \Delta P_D) = D\Delta f + (2H)\frac{d}{dt}\Delta f$$

$$0 - \frac{-40\,\text{MW}}{80\,\text{MW}} = 0 + (2)(5)\frac{d}{dt}\Delta f \Rightarrow \frac{d}{dt}\Delta f = 0.05 \frac{\text{pu}}{\text{s}} \Rightarrow$$

$$\frac{d}{dt}\Delta f = (0.05)(50 \text{ Hz}) = 2.5 \frac{\text{Hz}}{\text{s}} \Rightarrow$$

$$\Delta f = \left(2.5 \frac{\text{Hz}}{\text{s}}\right)(0.4\,\text{s}) = 1 \text{ Hz} \Rightarrow$$

$$f = f^0 + \Delta f = 50 + 1 = 51 \text{ Hz}$$

5.5 From equation 5.17, two generators' droops need to be in the same base. Assume that 500 MW is the base power.

$$R^{\text{new}} = R^{\text{old}} \left(\frac{S_b^{\text{new}}}{S_b^{\text{old}}} \right) \Rightarrow$$

$$\Rightarrow R_1^{\text{new}} = 0.02 \left(\frac{500}{200} \right) = 0.05, \; R_2^{\text{new}} = 0.03 \left(\frac{500}{600} \right) = 0.025$$

Then:

$$\Delta P_D = \frac{500\,\text{MW}}{S_b^{\text{new}}} = \frac{500}{500} = 1\,\text{pu}$$

From equations 5.14 and 5.15, we have:

$$\Delta f_{ss} = \frac{-\Delta P_D}{D_1 + D_2 + \dfrac{1}{R_1} + \dfrac{1}{R_2}} = \frac{-1}{0 + 0 + \dfrac{1}{0.05} + \dfrac{1}{0.025}} = -0.01667\,\text{pu Hz}$$

From equations 5.18–5.20, we have:

$$\Delta P_{T1} = \left(-\frac{\Delta f_{ss}}{R_1} \right) = \frac{0.01667}{0.05} = 0.3333\,\text{pu} = (0.3333)(500) = 166.67\,\text{MW}$$

$$\Delta P_{T2} = \left(-\frac{\Delta f_{ss}}{R_2} \right) = \frac{0.01667}{0.025} = 0.6667\,\text{pu} = (0.6667)(500) = 333.33\,\text{MW}$$

Checked: $P_{\text{load}} = 166.67 + 333.33 = 500\,\text{MW}$

5.6 We have the same answer as 5.5:

$$R^{\text{new}} = R^{\text{old}} \left(\frac{S_b^{\text{new}}}{S_b^{\text{old}}} \right) \Rightarrow R_3^{\text{new}} = 0.05 \left(\frac{500}{250} \right) = 0.1$$

$$\Delta f_{ss} = \frac{-\Delta P_D}{D + \dfrac{1}{R_1} + \dfrac{1}{R_2} + \dfrac{1}{R_3}} = \frac{-1}{0 + \dfrac{1}{0.05} + \dfrac{1}{0.025} + \dfrac{1}{0.1}} = -0.01429\,\text{pu Hz}$$

$$\Delta P_{T1} = \left(-\frac{\Delta f_{ss}}{R_1} \right) = \frac{0.01429}{0.05} = 0.2857\,\text{pu} = (0.2857)(500) = 142.86\,\text{MW}$$

$$\Delta P_{T2} = \left(-\frac{\Delta f_{ss}}{R_2} \right) = \frac{0.01429}{0.025} = 0.5714\,\text{pu} = (0.5714)(500) = 285.71\,\text{MW}$$

$$\Delta P_{T3} = \left(-\frac{\Delta f_{ss}}{R_3} \right) = \frac{0.01429}{0.1} = 0.1429 \text{ pu} = (0.1429)(500) = 71.43 \text{ MW}$$

Checked:

$$P_{load} = 142.86 + 285.71 + 71.43 = 500 \text{ MW}$$

5.7 We have the same answer as 5.5:

$$R_1^{new} = 0.05, \ R_2^{new} = 0.025, \ \Delta P_D = 1 \text{ pu}$$

From equation 5.17, we have:

$$D^{new} = D^{old} \left(\frac{S_b^{old}}{S_b^{new}} \right) \Rightarrow D_1^{new} = 10 \left(\frac{200}{500} \right) = 4, \ D_2^{new} = 5 \left(\frac{600}{500} \right) = 6$$

From equations 5.14 and 5.15, we have:

$$\Delta f_{ss} = \frac{-\Delta P_D}{D_1 + D_2 + \dfrac{1}{R_1} + \dfrac{1}{R_2}} = \frac{-1}{4 + 6 + \dfrac{1}{0.05} + \dfrac{1}{0.025}} = -0.0143 \text{ pu Hz}$$

Then:

$$\Delta P_{T1} = \left(-\frac{\Delta f_{ss}}{R_1} \right) = \frac{0.0143}{0.05} = 0.2860 \text{ pu} = (0.2860)(500) = 143 \text{ MW}$$

$$\Delta P_{T2} = \left(-\frac{\Delta f_{ss}}{R_2} \right) = \frac{0.0143}{0.025} = 0.5720 \text{ pu} = (0.5720)(500) = 286 \text{ MW}$$

$$\left(-\Delta f_{ss} D_1 \right) = 0.0143 \times 4 = 0.0572 \text{ pu} = (0.0572)(500) = 28.6 \text{ MW}$$

$$\left(-\Delta f_{ss} D_2 \right) = 0.0143 \times 6 = 0.0858 \text{ pu} = (0.0858)(500) = 42.9 \text{ MW}$$

Checked:

$$143 + 286 + 28.6 + 42.9 = 500.5 \approx 500 \text{ MW}$$

5.8 We have: $\Delta P_D = \dfrac{500 \text{ MW}}{S_b^{new}} = \dfrac{500}{500} = 1 \text{ pu}$

From equation 5.24:

$$\Delta f_{ss} = \frac{-\Delta P_{D1} - \cdots - \Delta P_{Dn}}{\left(D_1 + \dfrac{1}{R_1} \right) + \cdots + \left(D_n + \dfrac{1}{R_n} \right)} = \frac{-\Delta P_{D1}}{B_1 + \cdots + B_n} \Rightarrow$$

$$\Delta f_{ss} = \frac{-1}{\left(10+\dfrac{1}{0.02}\right)+\left(5+\dfrac{1}{0.03}\right)} = \frac{-1}{60+38.33} = -0.01017 \text{ pu Hz}$$

From equation 5.25:

$$\Delta P_{Line(i)} = \Delta P_{i1} + \cdots + \Delta P_{in} = -B_i \Delta f - \Delta P_{Di}, \ \Delta P_{ii} = 0 \Rightarrow$$

$$\Delta P_{Line(1)} = \Delta P_{12} = -B_1 \Delta f - \Delta P_{D1} = -(60)(-0.01017) - 1 = -0.3898 \text{ pu}$$

$$\Delta P_{12} = -(0.3898)(500) = -194.92 \text{ MW} = -\Delta P_{21}$$

From equations 5.18–5.20, we have:

$$\Delta P_{T1} = \left(-\frac{\Delta f_{ss}}{R_1}\right) = \frac{0.01017}{0.02} = 0.5085\,\text{pu} = (0.5085)(500) = 254.24 \text{ MW}$$

$$\Delta P_{T2} = \left(-\frac{\Delta f_{ss}}{R_2}\right) = \frac{0.01017}{0.03} = 0.3390\,\text{pu} = (0.3390)(500) = 169.5 \text{ MW}$$

$$\left(-\Delta f_{ss} D_1\right) = (0.01017)(10) = 0.1017\,\text{pu} = (0.1017)(500) = 50.85 \text{ MW}$$

$$\left(-\Delta f_{ss} D_2\right) = (0.01017)(5) = 0.05085\,\text{pu} = (0.05085)(500) = 25.425 \text{ MW}$$

Checked:

$$\text{Area 1: } \Delta P_{T1} + (-\Delta f_{ss} D_1) + \Delta P_{21} = \Delta P_{D1} \rightarrow$$

$$254.24 + 50.85 + 194.92 = 500.01 \approx 500 \text{ MW} = \Delta P_{D1}$$

Area 2: $\Delta P_{T2} + (-\Delta f_{ss} D_2) = \Delta P_{21} \rightarrow 169.5 + 25.425 = 194.93 \approx 194.92 \text{ MW} = \Delta P_{21}$

5.9 Area 2 does not interfere with the load change in Area 1 if the line between the two networks is opened. We have:

$$\Delta f_{ss} = \frac{-1}{\left(10+\dfrac{1}{0.02}\right)+(0)} = \frac{-1}{60} = -0.01667 \text{ pu Hz}$$

$$\Rightarrow (0.01667 \text{ pu Hz}) > (0.01017 \text{ pu Hz})$$

$$\Delta P_{T1} = \left(-\frac{\Delta f_{ss}}{R_1}\right) = \frac{0.01667}{0.02} = 0.8333 \text{ pu} = (0.8333)(500) = 416.67 \text{ MW}$$

$$\left(-\Delta f_{ss} D_1\right) = (0.01667)(10) = 0.1667 \text{ pu} = (0.1667)(500) = 83.35 \text{ MW}$$

Checked:

$$416.67 + 83.35 = 500.02 \approx 500 \text{ MW}$$

5.10 If the first generator's control loop is opened, we have:

$$\frac{1}{R_1} = 0 \Rightarrow$$

From answer 5.5, we have:

$$\Delta f_{ss} = \frac{-\Delta P_D}{D_1 + D_2 + \dfrac{1}{R_1} + \dfrac{1}{R_2}} = \frac{-1}{0 + 0 + (0) + \dfrac{1}{0.025}} = -0.025 \text{ pu Hz}$$

$$\Delta P_{T2} = \left(-\frac{\Delta f_{ss}}{R_2}\right) = \frac{0.025}{0.025} = 1 \text{ pu} = (1)(500) = 500 \text{ MW}$$

If the first generator's control loop is opened, the frequency drops more.

5.11 The integrator's input needs to be zero in the steady-state mode. From Figure 5.8, we have: $\Delta P_{Line1} = \Delta P_{12}$

In steady state ($s = 0$ or $t \to \infty$), then: $\Delta f_1 = \Delta f_2 = \Delta f$

From Figure 5.9, we have:

$$B_1 \, \Delta f + \Delta P_{12} = 0 \text{ and } B_2 \, \Delta f + \Delta P_{21} = 0$$

$$\Delta P_{12} = -\Delta P_{21} \to (B_1 \, \Delta f + \Delta P_{12}) + (B_2 \, \Delta f + \Delta P_{21}) = 0 \to (\Delta f = 0 \text{ and } \Delta P_{21} = 0)$$

From Figure 5.9, we have:

$$\left(\Delta P_{ref1} - \Delta P_{D1}\right)\left(k_p\right) = \Delta f = 0 \Rightarrow \Delta P_{ref1} = \Delta P_{D1}$$

5.12 We have:

$$\Delta P_{D1} = \frac{500 \text{ MW}}{S_b^{new}} = \frac{500}{500} = 1 \text{ pu}$$

If the line between two grids is opened, we have:

$$\Delta f_1 = \frac{59 - 60}{60} = -0.0167 \text{ pu}$$

$$\Delta f_{ss} = -0.01667 = \frac{-1}{\left(10 + \dfrac{1}{R_1}\right)} \Rightarrow R_1 = 0.02$$

$$\Rightarrow B_1 = D_1 + \frac{1}{R_1} = 10 + \frac{1}{0.02} = 60$$

If the line between two grids is closed, we have:

$$\Delta P_{Line(1)} = \Delta P_{12} = \frac{-195}{500} = -0.39 \text{ pu}$$

$$\Delta P_{12} = -B_1 \Delta f - \Delta P_{D1} \Rightarrow -0.39 = -(60)(\Delta f) - 1 \Rightarrow \Delta f = -0.01017 \text{ pu}$$

$$\Rightarrow \Delta f_{ss} = -0.01017 = \frac{-1}{\left(10 + \frac{1}{0.02}\right) + \left(5 + \frac{1}{R_2}\right)} \Rightarrow R_2 = 0.03$$

5.13 The open-loop AVR is:

$$G(s) = G_A(s) G_E(s) G_g(s) = \frac{K}{(1 + sT_A)(1 + sT_E)(1 + sT_g)} \Rightarrow$$

$$G(s) = \frac{K}{(1 + 0.1s)(1 + 0.5s)(1 + 5s)} = \frac{K}{0.25s^3 + 3.05s^2 + 5.6s + 1}$$

The characteristic equation is:

$$1 + G(s) = 0 \Rightarrow 1 + \frac{K}{0.25s^3 + 3.05s^2 + 5.6s + 1} = 0 \Rightarrow$$

$$0.25s^3 + 3.05s^2 + 5.6s + 1 + K = 0$$

The Routh–Hurwitz array is:

s^3	0.25	5.6
s^2	3.05	$1 + K$
s	$\dfrac{16.83 - 0.25K}{3.05}$	0
s^0	$1 + K$	0

From the s and s^0 rows we have:

$$\frac{16.83 - 0.25K}{3.05} > 0 \ \& \ 1 + K > 0 \Rightarrow -1 < K < 67.32$$

For $K = 67.32$, from the s^2 row we have:

$$3.05s^2 + 68.32 = 0 \Rightarrow s = \pm j4.733$$

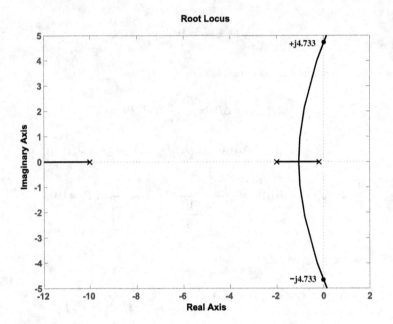

FIGURE 5.10   Root-locus diagram for answer 5.13.

To obtain the root-locus, we use the following commands in MATLAB.

rlocus(1, [0.25 3.05 5.6 1])

See Figure 5.10.

NOTE

1 In modern electric power plants, the original Watt governor has largely been replaced
   by more advanced and precise electronic control systems. These modern systems, such
   as proportional-integral-derivative (PID) controllers and other sophisticated automation technologies, provide much finer control over the speed and output of turbines and
   generators.

   However, the basic principle of speed regulation introduced by the Watt governor—
   using feedback to maintain consistent speed—remains foundational in these advanced
   systems. So while the Watt governor itself is no longer used, its legacy and fundamental
   concepts continue to influence modern control systems in power plants.

6 Key Concepts in Power System Analysis, Operation, and Control

Part One: Lesson Summary

6.1 INTRODUCTION

This chapter presents an overview of all the topics explored in this book, delving into advanced concepts related to analysis, operation, and control within power systems.

Professionals working in the field of power systems, especially those without a background in electrical engineering, will find this chapter highly beneficial.

6.2 OPTIMAL POWER FLOW (OPF)

In **load flow**, the magnitude and angle voltage of the buses are **unknown**, but the buses' power information is typically known (apart from the reference bus commonly called a **slack bus**) [1]. There are a lot of acceptable load flow solutions; because voltage changes are permitted within a range (±5%) of the nominal value [2,3]. **Optimal power flow** (OPF) [4] is a characteristic of one of the best load flows among many others.

Being "optimal" or "the best" could mean different things. In OPF, "better" can mean different things: "less losses, less costs, more income, more profit, more social welfare, less pollution, more security", etc. [5]. It is possible to turn everything above into money. Lagrange and Kuhn–Tucker methods can be used to solve the aforementioned problem [6].

All other calculations, including those for stability and short circuits, are based on load flow solutions [7–13]. For instance, numerous load flows are carried out in **unit commitment** and **contingency ranking** [14]. The well-known load flow techniques, Newton-Raphson, Gauss-Seidel, and Newton-Raphson-Seidel [15,16], were created for computing **voltage stability** [17,18] and figuring out the maximum transmission power [19–27]. Even though load flow calculations and their algorithms were among the first topics covered in the power system, the following factors have made load flow calculations more complex in the present day.

The significance of load flow calculations has also increased with distributed generation (DG) [28], FACTS devices [29], HVDC transmission lines [30–35], microgrid [36], solar [37], wind [38], biogas [39], electric vehicles [40], pumped-storage [41], state estimation [42], and smart grid [43–45].

DOI: 10.1201/9781003506751-6

Ultimately, it should be noted that new algorithms are still needed for the power system to solve load flow issues rapidly and accurately.

6.3   SHORT CIRCUIT

Planning and **operation** studies for power systems must include short circuit calculations [46]. Even though short circuit calculations can be made using load flow, they are usually the result of independent calculations. It is a normal task in power systems operation to compare the calculated short-circuit current with the maximum allowable value of switches. Power system planning also involves setting relays and determining the breaking power of circuit breakers [47].

Short-circuits to ground and symmetrical three-phase short-circuit currents are the most important ones to calculate. The symmetrical three-phase short circuit is the most important short circuit in buses far from the generator. One phase-to-ground short circuits are the most common short circuits in buses near to generators.

Fault current limiting reactors are usually used in the neutral of generators and earth star transformers to reduce single-phase short circuit current. The modification of the network structure, especially the open loop, is necessary to reduce the symmetrical three-phase short-circuit current.

6.4   APPROACHES TO ANALYZING POWER SYSTEM STABILITY

A power system should adequately accommodate the diverse power needs of both sizable and smaller consumers of domestic, commercial, and industrial sectors. It needs to endure the unpredictable impacts of nature with a reasonable level of reliability. In an era marked by elevated energy expenses, it is tasked with converting all primary available energy resources into electric power with the highest possible efficiency.

Within power systems, the control functions encompass a wide range of tasks. Certain control and decision-making processes, such as the efficient management of power flow, entail dynamics with very long time constants. Conversely, phenomena like the transients occurring on transmission lines due to lightning strikes should be handled within milliseconds. Computer-assisted human operators typically manage the slower control processes, while the faster control functions are entrusted to fully automatic control systems, which can be either open or closed-loop in nature [48].

A power system operates in its steady-state conditions if the following circumstances are met:

- All the load demands are met, and the Power Flow Equations (PFE) are satisfied.
- The frequency, f, is constant (60 Hz in the US, 50 Hz in the EU).
- The bus voltage magnitudes are within the prescribed limits.
- No component is overloaded.

Stability within a power system is essential to prevent disruptions, blackouts, or equipment failures that could lead to significant power outages.

This field encompasses various aspects, including steady-state stability, transient stability, and dynamic stability, each crucial in ensuring the system's resilience against fluctuations and disturbances. Understanding and analyzing power system stability involve assessing the system's ability to remain synchronized and stable under various conditions, considering factors like load changes, faults, or sudden disruptions.

Control mechanisms, ranging from manual intervention by operators to advanced automated systems, play a pivotal role in maintaining stability. These systems are designed to manage different time-scale phenomena, from slower processes handled by operators to faster responses managed by automatic control systems.

The general definition of **system stability** encompasses its ability to sustain an equilibrium state during regular operations and restore an acceptable equilibrium condition after being subjected to a disturbance.

In the context of AC power systems, power system stability refers to a state where all synchronous machines operating in parallel within the power grid maintain their synchronization.

6.4.1 TYPES OF POWER SYSTEM STABILITY

For analysis purposes, stability conditions which must be taken into account, are indicated in Figure 6.1.

6.4.2 SMALL SIGNAL ANALYSIS

Here the interest is to maintain synchronism among all generating units during or following a small, gradual, and slow varying load change (disturbance). These small disturbances clearly cannot cause loss of synchronism unless the system is operating

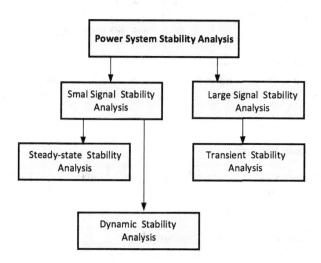

FIGURE 6.1 Forms of power system stability studies.

at or very near its stability limits threshold, due to the synchronizing torque developed by the generating units. For small changes in load or turbine speed, the system experiences **natural oscillations** in the generator torque angles, speeds, and EMFs. The system is stable if the amplitude of these oscillations is small and dampens out rapidly. The variations in rotor speeds of all synchronous machines cannot be uniform owing to large differences in their rotor inertia.

Small signal analysis can be subdivided into **steady-state stability** and **dynamic stability** analysis. The role of the **Automatic Voltage Regulator (AVR)** and **Automatic Load Frequency Control (ALFC)** is considered in the dynamic stability analysis domain. Generally, better damping can be achieved through the implementation of **Power System Stabilizers (PSS)**. Due to slow variations in the load, these studies are carried out for a considerably long time frame up to 30 seconds. Although the power system model is nonlinear, however, small signal analysis is carried out using the linearized mathematical model. This is because nonlinearities can be ignored when the study is carried out for long durations since a sufficient time frame for the system can inherently overcome all nonlinearities. The following definition is important.

6.4.3 STABILITY LIMIT DEFINITION

It pertains to the highest power level allowable for transfer to a designated location within the power system, given specific operating conditions, without risking the loss of synchronization (stability). The stability limit represents the peak electrical load the power system can handle while maintaining equilibrium, provided the energy input reaches a certain threshold value.

6.4.4 STEADY-STATE STABILITY

Steady-state stability can be defined as the ability of a power system to maintain synchronism among generating units within the system and external tie-lines, for small and slow natural load changes. The following definition is important:

6.4.4.1 Steady-State Stability Limit

The steady-state stability limit is the maximum power that can be transmitted to a point without loss of synchronism when the imposed disturbance is a slow, sustained, and small increase in the load.

6.4.5 DYNAMIC STABILITY

Dynamic stability can be described as the power system's capacity to retain synchronization following the initial swing until it returns to its newly established steady-state equilibrium state.

6.4.5.1 Definition of Dynamic Stability Limit

If the increase in the generator field current or regulations in speed settings occur simultaneously with an increase of load from the use of AVR and ALFC, the margin of stability limit will be significantly increased. The limit under these conditions is called the dynamic stability limit.

6.4.6 LARGE SIGNAL ANALYSIS—TRANSIENT STABILITY

In this type of analysis, the system is tested and analyzed for its stability following a large and sudden disturbance. Disturbances such as short circuits, clearing of faults, sudden changes in load, or loss of lines (opening) fall into such analysis. Owing to the severity of the disturbance, the analysis is carried out for a short time frame of up to 1 second. In this short duration, the system cannot overcome nonlinearities: the system is not behaving linearly, hence nonlinear mathematical models need to be considered for such an analysis. As long as AVR and ALFC are too slow to respond (in such a time frame) because of their time constants, therefore, they are not included in this study.

6.4.6.1 Transient Stability

Transient stability can be defined as the ability of the system to remain in synchronism during a period following a large and sudden disturbance and before the time when ALFC can act. Equal Area Criterion (EAC) is the graphical interpretation of the transient stability of the system. This method is only applicable to a single machine connected to an infinite bus system or a two-machine system. The concept of EAC is derived from the fact that the stored kinetic energy in the rotating mass tries to substantiate the imbalance between the machine output and input.

6.4.6.1.1 Definition of Transient Stability Limit

The maximum power that can be transmitted through the system without loss of stability under a sudden and large disturbance is referred to as the transient stability limit.

Part Two: Answer-Question

6.5 TWO-CHOICE QUESTIONS (YES/NO)

1. OPF is a characteristic of one of the best load flows among many others.
2. "More social welfare" is one of the meanings of "optimal" in OPF.
3. "Less pollution" is one of the meanings of "optimal" in OPF.
4. Load flow is the main component of unit commitment and contingency ranking.
5. Short circuit calculations can be made using load flow.
6. Power system planning involves setting relays and determining the breaking power of circuit breakers.
7. The symmetrical three-phase short circuit is the most important short circuit in buses far from the generator.
8. One-phase-to-ground short circuits are the most common short circuits in buses near generators.

9. Does power system stability refer to the ability of an electrical grid to maintain synchronous operation and equilibrium despite disturbances?
10. Are steady-state stability, transient stability, and dynamic stability different aspects that contribute to power system stability?
11. Can sudden load changes or faults impact power system stability?
12. Do control mechanisms, including both manual and automated systems, play a vital role in maintaining power system stability?
13. Are factors such as system inertia, damping, and control response time important in determining power system stability?
14. Is synchronism preservation a critical factor in ensuring power system stability?
15. Are tools like computer-based simulations and mathematical modeling commonly used to analyze power system stability?
16. Are stability limits determined to prevent the system from operating beyond safe parameters?
17. Can the integration of renewable energy sources present challenges for power system stability due to their intermittent nature?
18. Does the interaction between various system components, such as generators, transformers, and transmission lines, influence power system stability?
19. Steady-state stability of a power system is the ability of the power system to maintain synchronism among machines and on external tie lines.

6.6   ANSWERS TO TWO-CHOICE QUESTIONS

1 to 19	Yes

6.7   DESCRIPTIVE QUESTIONS

1. Define a slack bus.
2. Explain the Newton-Raphson-Seidel method.
3. Explain the different meanings of "optimal" in OPF.
4. Explain contingency ranking.
5. Discuss the types of short circuits.
6. Explain the difference between planning and operation studies.
7. Explain important differences among steady-state, transient-state, and dynamic stability.
8. Define power system stability and stability limit.
9. Discuss the methods to improve steady-state stability.
10. Discuss the methods to improve transient stability.

Appendix A
Complex Numbers

Phasor Representation

$$\hat{V} = V \angle \theta_V, \quad \hat{I} = I \angle \theta_I$$

Definition

$$Z = R + jX = |Z| \angle \theta, \quad |Z| = \sqrt{R^2 + X^2}$$

$$\theta = \begin{cases} R > 0, X > 0 \Rightarrow \theta = \tan^{-1}\left(\dfrac{X}{R}\right) \\[2mm] R < 0, X > 0 \Rightarrow \theta = 180 + \tan^{-1}\left(\dfrac{X}{R}\right) \\[2mm] R > 0, X < 0 \Rightarrow \theta = \tan^{-1}\left(\dfrac{X}{R}\right) \\[2mm] R < 0, X < 0 \Rightarrow \theta = -180 + \tan^{-1}\left(\dfrac{X}{R}\right) \end{cases}$$

Example A1

$$Z = \pm 3 \pm j4 \Rightarrow |Z| = \sqrt{(\pm 3)^2 + (\pm 4)^2} = 5$$

$$\theta = \begin{cases} R = 3, X = 4 \Rightarrow \theta = \tan^{-1}\left(\dfrac{4}{3}\right) = 53.13° \\[2mm] R = -3, X = 4 \Rightarrow \theta = 180 + \tan^{-1}\left(\dfrac{4}{-3}\right) = 180 - 53.13 = 126.87° \\[2mm] R = 3, X = -4 \Rightarrow \theta = \tan^{-1}\left(\dfrac{-4}{3}\right) = -53.13° \\[2mm] R = -3, X = -4 \Rightarrow \theta = -180 + \tan^{-1}\left(\dfrac{-4}{-3}\right) = -180 + 53.13 = -126.87° \end{cases}$$

Parallel Impedance:

$$Z = Z_1 \parallel Z_2 = \left(R_1 + jX_1\right) \parallel \left(R_2 + jX_2\right) = \frac{\left(R_1 + jX_1\right)\left(R_2 + jX_2\right)}{\left(R_1 + jX_1\right) + \left(R_2 + jX_2\right)} \Rightarrow$$

$$Z = \frac{\left(R_1 R_2 - X_1 X_2\right) + j\left(R_1 X_2 + R_2 X_1\right)}{\left(R_1 + R_2\right) + j\left(X_1 + X_2\right)} \triangleq \frac{Z_A}{Z_B} = \frac{|Z_A| \angle \theta_A}{|Z_B| \angle \theta_B} \Rightarrow$$

$$Z = \frac{|Z_A|}{|Z_B|} \angle (\theta_A - \theta_B)$$

Appendix B
Mathematical Formulas

$$\text{Cos}(53°) = \text{Sin}(37°) = 0.6$$

$$\text{Sin}(53°) = \text{Cos}(37°) = 0.8$$

$$\tan(37°) = 0.75, \ \tan(53°) = 1.333$$

$$1\angle\alpha + 1\angle(\alpha - 120) + 1\angle(\alpha + 120) = 0$$

$$1\angle\alpha + 1\angle(\alpha + 60) = \sqrt{3}\angle(\alpha + 30)$$

$$1\angle(\alpha + 60) - 1\angle\alpha + = 1\angle(\alpha + 120)$$

References

[1] Eidiani, M., Heidari, V., *Fundamentals of Power Systems Analysis 1: Problems and Solutions*, Taylor & Francis Group, CRC Press, pp. 1–215, https://doi.org/10.1201/9781003394433, 2023.

[2] Eidiani, M., "An Efficient Differential Equation Load Flow Method to Assess Dynamic Available Transfer Capability with Wind Farms," *IET Renewable Power Generation*, 15(16), pp. 3843–3855, https://doi.org/10.1049/rpg2.12299, 2021.

[3] Eidiani, M., "A New Load Flow Method to Assess the Static Available Transfer Capability," *Journal of Electrical Engineering and Technology*, 17(5), pp. 2693–2701, https://doi.org/10.1007/s42835-022-01105-3, 2022.

[4] Zeynal, H., Jiazhen, Y., Azzopardi, B., Eidiani, M., "Flexible Economic Load Dispatch Integrating Electric Vehicles," *2014 IEEE 8th International Power Engineering and Optimization Conference (PEOCO2014)*, 2014, pp. 520–525, https://doi.org/10.1109/PEOCO.2014.6814484.

[5] Zeynal, H., Zadeh, A.K., Nor, K.M., Eidiani, M., "Locational Marginal Price (LMP) Assessment Using Hybrid Active and Reactive Cost Minimization," *International Review of Electrical Engineering*, 5(5), pp. 2413–2418, 2010.

[6] Eidiani, M., Zeynal, H., Shaaban, M., "A Detailed Study on Prevailing ATC Methods for Optimal Solution Development," *2022 IEEE International Conference on Power and Energy (PECon)*, 2022, pp. 299–303, https://doi.org/10.1109/PECon54459.2022.9988775.

[7] Eidiani, M., "A Reliable and Efficient Method for Assessing Voltage Stability in Transmission and Distribution Networks," *International Journal of Electrical Power and Energy Systems*, 33(3), pp. 453–456, https://doi.org/10.1016/j.ijepes.2010.10.007, 2011.

[8] Eidiani, M., "A New Method for Assessment of Voltage Stability in Transmission and Distribution Networks," *International Review of Electrical Engineering*, 5(1), pp. 234–240, 2010.

[9] Eidiani, M., Ashkhane, Y., Khederzadeh, M., "Reactive Power Compensation in Order to Improve Static Voltage Stability in a Network with Wind Generation," *2009 International Conference on Computer and Electrical Engineering, ICCEE 2009*, 2009, pp. 47–50, vol. 1, https://doi.org/10.1109/ICCEE.2009.239.

[10] Eidiani, M., Buygi, M.O., Ahmadi, S., "CTV, Complex Transient and Voltage Stability: A New Method for Computing Dynamic ATC," *International Journal of Power and Energy Systems*, 26(3), pp. 296–304, https://doi.org/10.2316/Journal.203.2006.3.203-3597, 2006.

[11] Eidiani, M., Badokhty, M.E., Ghamat, M., Zeynal, H., "Improving Transient Stability Using Combined Generator Tripping and Braking Resistor Approach," *International Review on Modelling and Simulations*, 4(4), pp. 1690–1699, 2011.

[12] Eidiani, M., Baydokhty, M.E., Ghamat, M., Zeynal, H., Mortazavi, H., "Transient Stability Enhancement via Hybrid Technical Approach," *2011 IEEE Student Conference on Research and Development*, 2011, pp. 375–380, https://doi.org/10.1109/SCOReD.2011.6148768.

[13] Eidiani, M., Shanechi, M.H.M., Vaahedi, E., "Fast and Accurate Method for Computing FCTTC (First Contingency Total Transfer Capability)," *Proceedings. International Conference on Power System Technology*, 2002, pp. 1213–1217, vol. 2, https://doi.org/10.1109/ICPST.2002.1047595.

[14] Zeynal, H., Hui, L.X., Jiazhen, Y., Eidiani, M., Azzopardi, B., "Improving Lagrangian Relaxation Unit Commitment with Cuckoo Search Algorithm," *2014 IEEE International Conference on Power and Energy (PECon)*, 2014, pp. 77–82, https://doi.org/10.1109/PECON.2014.7062417.

[15] Eidiani, M., Zeynal, H., Zadeh, A.K., Mansoorzadeh, S., Nor, K.M., "Voltage Stability Assessment: An Approach with Expanded Newton Raphson-Sydel," *2011 5th International Power Engineering and Optimization Conference*, 2011, pp. 31–35, https://doi.org/10.1109/PEOCO.2011.5970424.

[16] Eidiani, M., "Assessment of Voltage Stability with New NRS," *2008 IEEE 2nd International Power and Energy Conference*, 2008, pp. 494–496, https://doi.org/10.1109/PECON.2008.4762525.

[17] Eidiani, M., Yazdanpanah, D., "Minimum Distance, A Quick and Simple Method of Determining the Static ATC," *Journal of Electrical Engineering*, 11(2), pp. 95–101, 2011.

[18] Eidiani, M., Asadi, S.M., Faroji, S. A., Velayati, M.H., Yazdanpanah, D., "Minimum Distance, A Quick and Simple Method of Determining the Static ATC," *2008 IEEE 2nd International Power and Energy Conference*, 2008, pp. 490–493, https://doi.org/10.1109/PECON.2008.4762524.

[19] Eidiani, M., "A Rapid State Estimation Method for Calculating Transmission Capacity Despite Cyber Security Concerns," *IET Generation, Transmission and Distribution*, 17(20), pp. 4480–4488, https://doi.org/10.1049/gtd2.12747, 2023.

[20] Eidiani, M., "A Reliable and Efficient Holomorphic Approach to Evaluate Dynamic Available Transfer Capability," *International Transactions on Electrical Energy Systems*, 31(11), p. e13031, https://doi.org/10.1002/2050-7038.13031, 2021.

[21] Eidiani, M., Zeynal, H., Zadeh, A.K., Nor, K.M., "Exact and Efficient Approach in Static Assessment of Available Transfer Capability (ATC)," *2010 IEEE International Conference on Power and Energy*, 2010, pp. 189–194, https://doi.org/10.1109/PECON.2010.5697580.

[22] Eidiani, M., Zeynal, H., "A Fast Holomorphic Method to Evaluate Available Transmission Capacity with Large Scale Wind Turbines," *9th Iranian Conference on Renewable Energy & Distributed Generation (ICREDG)*, 2022, pp. 1–5, https://doi.org/10.1109/ICREDG54199.2022.9804527.

[23] Eidiani, M., Zeynal, H., Zakaria, Z., "An Efficient Holomorphic Based Available Transfer Capability Solution in Presence of Large Scale Wind Farms," *2022 IEEE International Conference in Power Engineering Application (ICPEA)*, 2022, pp. 1–5, https://doi.org/10.1109/ICPEA53519.2022.9744711.

[24] Eidiani, M., Shanechi, M.H.M., "FAD-ATC: A New Method for Computing Dynamic ATC," *International Journal of Electrical Power and Energy Systems*, 28(2), pp. 109–118, https://doi.org/10.1016/j.ijepes.2005.11.004, 2006.

[25] Eidiani, M., "Atc Evaluation by CTSA and POMP, Two New Methods for Direct Analysis of Transient Stability," *IEEE/PES Transmission and Distribution Conference and Exhibition*, 2002, pp. 1524–1529, vol. 3, https://doi.org/10.1109/TDC.2002.1176824.

[26] Eidiani, M., Zeynal, H., Zakaria, Z., "Development of Online Dynamic ATC Calculation Integrating State Estimation," *2022 IEEE International Conference in Power Engineering Application (ICPEA)*, 2022, pp. 1–5, https://doi.org/10.1109/ICPEA53519.2022.9744694.

[27] Eidiani, M., Zeynal, H., Zakaria, Z., "An Efficient Method for Available Transfer Capability Calculation Considering Cyber-Attacks in Power Systems," *2023 IEEE 3rd International Conference in Power Engineering Applications (ICPEA)*, Putrajaya, Malaysia, 2023, pp. 127–130, https://doi.org/10.1109/ICPEA56918.2023.10093168.

[28] Eidiani, M., Ghavami, A., "New Network Design for Simultaneous Use of Electric Vehicles, Photovoltaic Generators, Wind Farms and Energy Storage," 2022 *9th Iranian Conference on Renewable Energy & Distributed Generation (ICREDG)*, pp. 1–5, https://doi.org/10.1109/ICREDG54199.2022.9804534, 2022.

[29] Eidiani, M., Zeynal, H., "An Effective Method to Determine the Available Transmission Capacity with Variable Frequency Transformer," *International Transactions on Electrical Energy Systems*, 2023, p. 8404284, https://doi.org/10.1155/2023/8404284, 2023.

[30] Eidiani, M., "A New Hybrid Method to Assess Available Transfer Capability in AC-DC Networks Using the Wind Power Plant Interconnection," *IEEE Systems Journal*, 17(1), pp. 1375–1382, https://doi.org/10.1109/JSYST.2022.3181099, 2023.

[31] Eidiani, M., Zeynal, H., Zakaria, Z., Shaaban, M., "A Comprehensive Study on the Renewable Energy Integration Using DIgSILENT," *2023 IEEE 3rd International Conference in Power Engineering Applications (ICPEA)* , Putrajaya, Malaysia, 2023, pp. 197–201, https://doi.org/10.1109/ICPEA56918.2023.10093153.

[32] Eidiani, M., Zeynal, H., Zakaria, Z., Shaaban, M., "Analysis of Optimization Methods Applied for Renewable Energy Integration," *2023 IEEE 3rd International Conference in Power Engineering Applications (ICPEA)*, March 2023, pp. 6–7.

[33] Eidiani, M., "Applying Optimization Techniques to Develop a Renewable Energy Supply Map," In: Fathi, M., Zio, E., Pardalos, P. M. (eds) *Handbook of Smart Energy Systems*. Springer, Cham. https://doi.org/10.1007/978-3-030-72322-4_61-1, 2022.

[34] Eidiani, M., "Integration of Renewable Energy Sources," In: Fathi, M., Zio, E., Pardalos, P. M. (eds) *Handbook of Smart Energy Systems*. Springer, Cham. https://doi.org/10.1007/978-3-030-72322-4_41-1, 2022.

[35] Eidiani, M., Zeynal, H., Zakaria, Z., "A Comprehensive Study on the Renewable Energy Integration Using DIgSILENT," *2023 IEEE 3rd International Conference in Power Engineering Applications (ICPEA)*, Putrajaya, Malaysia, 2023, pp. 197–201, https://doi.org/10.1109/ICPEA56918.2023.10093153.

[36] Eidiani, M., Kargar, M., "Frequency and Voltage Stability of the Microgrid with the Penetration of Renewable Sources," *2022 9th Iranian Conference on Renewable Energy & Distributed Generation (ICREDG)*, pp. 1–6, https://doi.org/10.1109/ICREDG54199.2022.9804542, 2022.

[37] Eidiani, M., Zeynal, H., Ghavami, A., Zakaria, Z., "Comparative Analysis of Mono-Facial and Bifacial Photovoltaic Modules for Practical Grid-Connected Solar Power Plant Using PVsyst," *2022 IEEE International Conference on Power and Energy (PECon)*, pp. 499–504, https://doi.org/10.1109/PECon54459.2022.9988872, 2022.

[38] Eidiani, M., Shahdehi, N.A., Zeynal, H., "Improving Dynamic Response of Wind Turbine Driven DFIG with Novel Approach," *2011 IEEE Student Conference on Research and Development*, 2011, pp. 386–390, https://doi.org/10.1109/SCOReD.2011.6148770.

[39] Eidiani, M., Mahnani, S., "Optimally and Independent Planned Microgrid with Solar-Wind and Biogas Hybrid Renewable Systems by HOMER," *Majlesi Journal of Electrical Engineering*, 17(2), pp. 153–158, https://doi.org/10.30486/mjee.2023.1974843.1021, 2023.

[40] Zeynal, H., Jiazhen, Y., Azzopardi, B., Eidiani, M., "Impact of Electric Vehicle's integration into the economic VAr dispatch algorithm," *2014 IEEE Innovative Smart Grid Technologies - Asia (ISGT ASIA)*, 2014, pp. 780–785, https://doi.org/10.1109/ISGT-Asia.2014.6873892.

[41] Zeynal, H., Eidiani, M., "Hydrothermal Scheduling Flexibility Enhancement with Pumped-Storage Units," *2014 22nd Iranian Conference on Electrical Engineering (ICEE)*, 2014, pp. 820–825, https://doi.org/10.1109/IranianCEE.2014.6999649.

[42] Eidiani, M., Zeynal, H., "Determination of Online DATC with Uncertainty and State Estimation," *2022 9th Iranian Conference on Renewable Energy & Distributed Generation (ICREDG)*, 2022, pp. 1–6, https://doi.org/10.1109/ICREDG54199.2022.9804581.

[43] Zeynal, H., Eidiani, M., Yazdanpanah, D., "Intelligent Substation Automation Systems for Robust Operation of Smart Grids," *2014 IEEE Innovative Smart Grid Technologies - Asia (ISGT ASIA)*, 2014, pp. 786–790, https://doi.org/10.1109/ISGT-Asia.2014.6873893.

[44] Zand, Z., Ghahri, M.R., Majidi, S., Eidiani, M., Nasab, M.A., Zand, M., "Smart Grid and Resilience," In: Fathi, M., Zio, E., Pardalos, P. M. (eds) *Handbook of Smart Energy Systems*. Springer, Cham. https://doi.org/10.1007/978-3-030-72322-4_178-1, 2022.

[45] Momen, S., Hekmati, A., Majidi, S., Zand, Z., Zand, M., Nikoukar, J., Eidiani, M., "Energy Harvesting for Smart Energy Systems," In: *Handbook of Smart Energy Systems*. Springer, Cham. https://doi.org/10.1007/978-3-030-72322-4_12-1, 2022.

[46] Baydokhty, M.E., Eidiani, M., Zeynal, H., Torkamani, H., Mortazavi, H., "Efficient Generator Tripping Approach with Minimum Generation Curtailment based on Fuzzy System Rotor Angle Prediction," *Przeglad Elektrotechniczny*, 88(9 A), pp. 266–271, 2012.

[47] Ghardashi, G., Gandomkar, M., Majidi, S., Eidiani, M., Dadfar, S., "Accuracy and Speed Improvement of Microgrid Islanding Detection based on PV using Frequency-Reactive Power Feedback Method," *2022 International Conference on Protection and Automation of Power Systems (IPAPS)*, 2022, pp. 1–8, https://doi.org/10.1109/IPAPS55380.2022.9763190.

[48] Eidiani, M., Zeynal, H., "New Approach Using Structure-Based Modeling for the Simulation of Real Power/Frequency Dynamics in Deregulated Power Systems," *Turkish Journal of Electrical Engineering and Computer Sciences*, 22(5), pp. 1130–1146, https://doi.org/10.3906/elk-1208-90, 2014.

Units, Symbols, Notations, Abbreviations

UNITS

A	Ampere
AC	Alternating current
cm	Centimeter
DC	Direct current
F	Farad
H	Henry
h	Hour
hp	Horsepower
HV	High voltage HV
Hz	Hertz
J	Joule
kV	Kilovolt
kVA	Kilovolt ampere
kW	Kilowatt
MW	Megawatt
MΩ	Megaohm
N	Newton
N.m	Newton meter
pu	Per unit
rad	Radian
s	Second (time)
V	Volt
VA	Volt-ampere
VAr	Volt-ampere reactive
W	Watt
Ω	Ohm

SYMBOLS

$\alpha i, \beta i, \gamma i$	Constant parameters related to the ith cost power plant
$\measuredangle$	Voltage angle
$\parallel$	Parallel sign
λ	Lagrange multiplier or Lagrange coefficient
Δ	Small changes
Aacc	Acceleration area
Adec	Deceleration area

Bi	Frequency bias factor or actual frequency response characteristic or Area bias
Bij	Kron's coefficients or B coefficients.
$C(P)$	Cost function
CB	Circuit breakers
D	Load sensitivity coefficient to the frequency
cr	Critical clearing
cl	Clearing
e	Electrical
Er	Total error
(f) or (F)	Fault
H	Generator inertia constant
inv or $(^{-1})$	Inverse
L	Lagrangian or Lagrange function
m	Mechanical
max	Maximum
min	Minimum
(o)	Old things
P_D	All the load demands
P_T	Turbine power
R	Governor's speed regulation or droop
ss	Steady-State
s.t.	such that or subject to
T_T	Turbine time constants
T_H	Hydraulic time constants
T_P	Loaded generator time constants
T_{ij}^0	Tie line synchronizing power coefficient
tot	Total

NOTATION

Lowercase letters such as $v(t)$ and $i(t)$ indicate instantaneous values.
Uppercase letters in italics such as V and I indicate **rms** values.
Uppercase letters such as $\hat{V}$ and indicate **rms** phasors.
Matrices and vectors with real components such as $\boldsymbol{R}$ and $\boldsymbol{I}$ are indicated by boldface type.
Matrices and vectors with complex components such as $\boldsymbol{Z}$ and $\boldsymbol{I}$ are indicated by boldface italic type.
Superscript (T) denotes vector or matrix transpose.
Asterisk (*) denotes complex conjugate.

ABBREVIATIONS

AFC	Automatic Frequency Control
ALFC	Automatic Load Frequency Control
AVR	Automatic Voltage Regulator

ACE	Area Control Error
DG	Distributed Generation
EAC	Equal Area Criterion
FACTS	Flexible Alternating Current Transmission System
GS	Gauss-Seidel
HVDC	High-Voltage Direct Current
ITL	Incremental Transmission Loss
KCL	Kirchoff's Current Law
KVL	Kirchhoff's Voltage Law
NR	Newton Raphson
NRS	Newton Raphson Seidel
OPF	Optimal Power Flow
PF	Power Flow
PFi	Penalty Factor of the i th line
PFE	Power Flow Equations
PSS	Power System Stabilizers
SCC	Short Circuit Capacity
SMES	Superconductive Magnetic Energy Storage

Bibliography

When selecting a book on **Power System Analysis**, you should consider your level of expertise, the depth of information you require, and whether you prefer a more theoretical- or application-oriented approach.

The following books can serve as valuable resources for students, engineers, and professionals looking to enhance their knowledge of power system analysis.

Below is a list of prominent books on power system analysis (listed without any specific order):

[1] Eidiani, M., Heidari, V., *Fundamentals of Power Systems Analysis 1: Problems and Solutions*, Taylor & Francis Group, CRC Press, 1st edition, Boca Raton, pp. 1–215, https://doi.org/10.1201/9781003394433, 2023.

[2] Elgerd, O. I., *Electric Energy Systems Theory*, Mc Graw Hill India, 2nd edition, 1998.

[3] Stevenson, W. D., *Elements of Power System Analysis Hardcover*, McGraw-Hill, 4th edition, 1982.

[4] Gross, C. A., *Power System Analysis*, Wiley, 2nd edition, 1986.

[5] Grainger, J. J., and Stevenson Jr., W. D., *Power System Analysis*, McGraw Hill, 1st edition, 1994.

[6] Glover, J. D., Overbye, T., Sarma, M. S., and Birchfield, A. B., *Power System Analysis and Design*, Cengage Learning, 7th edition, 2022.

[7] Saadat, H., *Power Systems Analysis*, PSA Publishing, 3rd edition, 2010.

[8] Gomez-Exposito, A., Conejo, A. J., Canizares, C., *Electric Energy Systems: Analysis and Operation*, CRC Press, Boca Raton, 2017.

[9] Kothari, D. P., Nagrath, I. J., Saket, R. K., *Modern Power System Analysis*, McGraw Hill, 5th Edition, 2022.

[10] Ramana, N. V. *Power System Operation and Control*, Pearson, 1st edition, 2010.

[11] Sivanagaraju, S., Sreenivasan, G., *Power System Operation and Control*, Pearson Education, 1st edition, 2012.

[12] Sodhi, R. *Simulation and Analysis of Modern Power Systems*, McGraw Hill, 1st edition, 2021.

[13] Bergen, A., Vittal, V., *Power Systems Analysis*, Pearson; 2nd edition, 1999.

Index

Printed in the United States
by Baker & Taylor Publisher Services